Martial Rose Library
Tel: 01962 827306

- 2 FEB 2012

1 3 MAR 2012

1 0 MAY 2012

1 2 MAY 2012

- 1 OCT 2012

1 2 NOV 2012

SEVEN DAY LOAN ITEM
To be returned on or before the day marked above, subject to recall.

Staging and Performing Translation

Also by Roger Baines

'INQUIÉTUDE' IN THE WORK OF PIERRE MAC ORLAN

Also by Manuela Perteghella

ONE POEM IN SEARCH OF A TRANSLATOR: Re-writing 'Les Fenêtres' by Apollinaire (*edited with Eugenia Loffredo*)

TRANSLATION AND CREATIVITY: Perspectives on Creative Writing and Translation Studies (*edited with Eugenia Loffredo*)

Staging and Performing Translation

Text and Theatre Practice

Edited by

Roger Baines
Lecturer in French and Translation Studies, University of East Anglia

Cristina Marinetti
Assistant Professor in Translation Studies, University of Warwick

Manuela Perteghella
Senior Lecturer in Translation Studies, London Metropolitan University

Selection and editorial matter © Roger Baines, Cristina Marinetti and Manuela Perteghella 2011
Chapters © their authors 2011

All rights reserved. No reproduction, copy or transmission of this publication may be made without written permission.

No portion of this publication may be reproduced, copied or transmitted save with written permission or in accordance with the provisions of the Copyright, Designs and Patents Act 1988, or under the terms of any licence permitting limited copying issued by the Copyright Licensing Agency, Saffron House, 6–10 Kirby Street, London EC1N 8TS.

Any person who does any unauthorized act in relation to this publication may be liable to criminal prosecution and civil claims for damages.

The authors have asserted their rights to be identified as the authors of this work in accordance with the Copyright, Designs and Patents Act 1988.

First published 2011 by
PALGRAVE MACMILLAN

Palgrave Macmillan in the UK is an imprint of Macmillan Publishers Limited, registered in England, company number 785998, of Houndmills, Basingstoke, Hampshire RG21 6XS.

Palgrave Macmillan in the US is a division of St Martin's Press LLC, 175 Fifth Avenue, New York, NY 10010.

Palgrave Macmillan is the global academic imprint of the above companies and has companies and representatives throughout the world.

Palgrave® and Macmillan® are registered trademarks in the United States, the United Kingdom, Europe and other countries.

ISBN: 978–0–230–22819–1   hardback

This book is printed on paper suitable for recycling and made from fully managed and sustained forest sources. Logging, pulping and manufacturing processes are expected to conform to the environmental regulations of the country of origin.

A catalogue record for this book is available from the British Library.

Library of Congress Cataloging-in-Publication Data

Staging and performing translation : text and theatre practice /
   edited by Roger Baines, Cristina Marinetti and Manuela Perteghella.
      p. cm.
   Includes bibliographical references and index.
   ISBN 978–0–230–22819–1 (alk. paper)
   1. Drama—Translating. 2. Translating and interpreting. 3. Theater.
     I. Baines, Roger W. II. Marinetti, Christina, 1976– III. Perteghella, Manuela.
PN886.S73 2011
792.1—dc22   2010033938

10   9   8   7   6   5   4   3   2   1
20 19 18 17 16 15 14 13 12 11

Printed and bound in Great Britain by
CPI Antony Rowe, Chippenham and Eastbourne

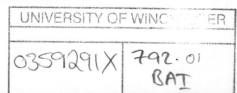

Contents

List of Figures and Table   vii

Acknowledgements   viii

Notes on Contributors   x

Introduction   1
Cristina Marinetti, Manuela Perteghella, Roger Baines

Part I   Explorations and Experiments in Theory and Practice   9

1. Metaphor and Metonymy: the Translator-Practitioner's Visibility   11
David Johnston

2. The Translator as *metteur en scène*, with Reference to *Les Aveugles* [*The Blind*] by Maurice Maeterlinck   31
Carole-Anne Upton

3. Musical Realizations: a Performance-based Translation of Rhythm in Koltès' *Dans la solitude des champs de coton*   49
Roger Baines and Fred Dalmasso

4. The Theatre Sign Language Interpreter and the Competing Visual Narrative: the Translation and Interpretation of Theatrical Texts into British Sign Language   72
Siobhán Rocks

5. Inferential Meaning in Drama Translation: the Role of Implicature in the Staging Process of Anouilh's *Antigone*   87
Alain J. E. Wolf

Part II   Practical Perspectives on Translation, Adapting and Staging   105

6. Translating Bodies: Strategies for Exploiting Embodied Knowledge in the Translation and Adaptation of Chinese *Xiqu* Plays   107
Megan Evans

7. Brecht's *The Threepenny Opera* for the National Theatre: a 3p Opera?   126
Anthony Meech

vi Contents

8 The Translator as Cultural Promoter: or how Renato
   Gabrielli's *Qualcosa Trilla* went on the Road as *Mobile
   Thriller*   139
   Margaret Rose and Cristina Marinetti

9 *Cow-boy poétré*: a Bilingual Performance for a Unilingual
   Audience   155
   Louise Ladouceur and Nicole Nolette

Part III   In Conversation with Practitioners   **171**

10 Interview with Christopher Hampton   173
   Roger Baines and Manuela Perteghella

11 Not Lost in Translation   187
   Jack Bradley

12 Roundtable on Collaborative Theatre Translation Projects:
   Experiences and Perspectives   200
   Jonathan Meth, Katherine Mendelsohn, Zoë Svendsen

Part IV   Politics, Ethics and Stage Translation   **213**

13 *The Impotents*: Conflictual Significance Imposed on the
   Creation and Reception of an Arabic-to-Hebrew
   Translation for the Stage   215
   Yotam Benshalom

14 The Politics of Translating Contemporary French Theatre:
   how 'Linguistic Translation' Becomes 'Stage Translation'   230
   Clare Finburgh

15 Translating Zapolska: Research through Practice   249
   Teresa Murjas

Index   267

List of Figures and Table

Figures

4.1 The interaction of the three areas concerning the theatre sign language interpreter — 76

4.2 The complex assignment (a model approach to sign language interpreting for the theatre) — 79

6.1 Author's rehearsal script of Elizabeth Wichmann-Walczak's working translation of the traditional *jingju* play *Judge Bao and the Case of Qin Xianglian* (Shen Xiaomei et al., 2002), performed at Kennedy Theatre, University of Hawai'i at Mānoa, February 2002 — 115

7.1 Elizabeth Marsh singing 'Pirate Jenny's Song' in *The Threepenny Opera* (*Die Dreigroschenoper*), National Theatre, London, February 2003 — 136

Table

9.1 Language use in various versions of the play *Cow-boy poétré* — 166

Acknowledgements

Firstly the editors would like to record their gratitude to the following: all our contributors for their cooperation in drawing this collection together; Paula Kennedy, Ben Doyle and Christine Ranft from Palgrave Macmillan for their enthusiasm and expertise, Carl Lavery for insightful comments on the organization and content of the manuscript; and Kaye Mackay and Jon Cook from the University of East Anglia for their assistance in initiating the project from which this volume has come.

Secondly, the editors would like to thank the following publishers for granting permission to reproduce extracts from the works listed below:

Editions de la Table Ronde for Jean Anouilh's *Antigone* (1946)

AC Black publishers for 'In the Solitude of Cotton Fields' by Bernard-Marie Koltès translated by Jeffrey Wainwright (2004), for *Sing Yer Heart Out for the Lads* by Roy Williams (2002), and for *Antigone* by Jean Anouilh translated by Barbara Bray (2000)

Les Editions de Minuit for Bernard-Marie Koltès' *Dans la solitude des champs de coton* (2004)

University Press of America for 'Injustice Done to Dou E' by Guan Hanqing translated by Qian Ma (2005)

Plays International Company for *Mobile Thriller* by Renato Gabrielli translated by Margaret Rose (2004)

Hakim Printing and Publishing for *The Impotents* by Riad Masarwy (2001)

Théâtre Ouvert for *Par les routes* by Noëlle Renaude (2005)

Duckworth Publishers for *Pelléas et Mélisande, Les Aveugles, L'Intruse, L'Intérieur* by Maurice Maeterlinck (edited by Leighton Hodson, 1999)

Bill Kosmas for *Bodas de sangre* by Federico Lorca translated by G. Lujan/ R. O'Connell (Penguin, 1961), Ted Hughes (Faber, 1996) and Brendan Kennelly (Bloodaxe, 1996)

Seamus Heaney for his poem 'Peter Street at Bankside'

Finally, the editors would like to also thank the following for permission to reproduce images: Josh Barnes for the cover image from *Women Generals of the Yang Family* (translated by Elizabeth Wichmann-Walczak with Hui-mei Chang) at the Kennedy Theatre, University of Hawai'i at Mānoa (USA), 2006; Daniel Cournoyer, artistic and managing director of L'Unithéâtre, for the cover image from *Cow-boy Poétré* by Kenneth Brown at L'Unithéâtre, Edmonton (Canada), 2005; and Lib Taylor for the cover image from *Miss Maliczewska* by Gabriela Zapolska (translated and directed by Teresa Murjas), University of Reading/POSK Theatre, London (UK), 2008. Chapter 7's image from *The Threepenny Opera* by Bertolt Brecht (translation Jeremy Sams and Anthony Meech) at the National Theatre, 2003, is by the late Ivan Kyncl.

The publishers wish to state that they have made every effort to contact the copyright holders, but if any have been overlooked the publishers will be pleased to make the necessary arrangement at the first opportunity.

Notes on Contributors

Roger Baines is currently Head of the School of Language and Communication Studies at the University of East Anglia (UK) where he teaches and researches Translation Studies and French language. He has published on the work of Pierre Mac Orlan, personal insults and gender, and ritual insults, in contemporary French, and the translation and adaptation of Adel Hakim's 1990 play *Exécuteur 14*. He is co-host and co-founder of the STRAP (Stage Translation Research Adaptation Practice) discussion list.

Yotam Benshalom lives in Jerusalem and translates drama and poetry between Hebrew, Arabic and English. He is completing a Ph.D. about the relevance and applicability of acting and performance theories to the practice of translation at the University of Warwick (UK). His interests include metaphors of translation, poetics of translation and translation ethics.

Jack Bradley began work as a playwright in 1975 (*Stepping Stones*, Royal Court, London, Young Writers' Festival) and continued to do so throughout the 80s and early 90s, with twenty productions to his name. Over time, he became more involved in play development and literary management and worked at the Soho Theatre, London (1989–94) before joining the Royal National Theatre of Great Britain where he was Literary Manager for twelve years, advising on the repertoire for Richard Eyre, Trevor Nunn and Nicholas Hytner, spearheading their new play policy. As a dramaturg he has taught and run workshops internationally – from Belfast to Buenos Aires, Oslo to Soweto. He has lectured and taught on creative courses throughout Britain. He is now a freelance dramaturg and has resumed work as a playwright and translator. In addition, he is Literary Associate to Sonia Friedman Productions in the West End of London, Associate of the Tricycle Theatre, London, and advisor to the Scottish Arts Council. He is currently visiting Lecturer to Newcastle University (UK) and is writing a book on playwriting for Oberon Books.

Fred Dalmasso is studying for a Ph.D. on French philosopher Alain Badiou's theatre at the University of Loughborough (UK) and is working on practice-based theatre translation in collaboration with Roger Baines and poet David 'Stickman' Higgins. He has published on the translation and adaptation of Adel Hakim's 1990 play *Exécuteur 14*. He has also worked as an actor and theatre director.

Megan Evans teaches Theatre at Victoria University of Wellington, New Zealand. She has published on directing in *xiqu* (Chinese opera) and interactions between *xiqu* for stage and moving image media. She trained in *xiqu* performance at the University of Hawai'i at Mānoa and the Academy of Chinese Theatre Arts in Beijing. Recent directing credits include adaptations of Guan Hanqing's *Injustice Done to Dou'E* and Anthony C. Yu's translation of the classic Chinese novel *Journey to the West*.

Clare Finburgh teaches Modern Drama and European Literature at the University of Essex (UK). The main focus of her research is the theatre of Jean Genet, on whom she has published many articles, co-edited a volume of essays, *Genet: Performance and Politics* (2006), and is currently co-authoring a book for the Routledge Modern and Contemporary Dramatists collection. In addition, she is co-editing a collection of essays on contemporary French theatre and performance (Palgrave Macmillan, 2010). Other areas of research include post-9/11 British theatre, on which she has published widely in France. In 2007, she translated into English Noëlle Renaude's *Par les routes*, which was performed at the Edinburgh Fringe Festival.

David Johnston is Head of the School of Languages, Literatures and Performing Arts at Queen's University Belfast (UK). Among his various publications on translation is *Stages of Translation* (1996), which set down the parameters for discussion of translation for performance, and recently, with Stephen Kelly, *Betwixt and Between: Place and Cultural Translation* (2007). He is an award-winning translator for the stage who has been commissioned three times by the Royal Shakespeare Company, and has worked regularly with the BBC and London's Royal Court Theatre. His translation of Lope de Vega's *The Dog in the Manger* recently had a ten-week run at the Washington Shakespeare Theatre Company.

Louise Ladouceur is Professor at the University of Alberta's Campus Saint-Jean (Canada), and Associate French Editor of *Theatre Research in Canada*. Her research focuses on theatre translation, Canadian drama and the francophone drama repertoire of Western Canada. Her book *Making the Scene: la traduction du théâtre d'une langue officielle à l'autre au Canada* (2005), winner of the Gabrielle Roy Prize and the Ann Saddlemyer Book Award, will be published in English in 2011 by the University of Alberta Press.

Cristina Marinetti is Assistant Professor in Translation Studies at the University of Warwick (UK) and Associate Editor of the translation journal *Target*. She has published on translation theory, translation for performance, Italian and British theatre, with a specialization in Goldoni and *commedia dell'arte*. Recent projects explore meeting grounds between translation and performance in multilingual, migrant theatres. She is co-host and co-founder of the STRAP (Stage Translation Research Adaptation Practice) discussion list.

Anthony Meech teaches in the Department of Drama and Music at the University of Hull (UK). He has published on German Theatre from the eighteenth century to the present day, as well as translating numerous plays from Lenz and Büchner to Brecht, and the work of contemporary German dramatists.

Katherine Mendelsohn is the Literary Manager at the Traverse Theatre, Edinburgh, Scotland. In 2000, Katherine launched the Traverse Theatre's *Playwrights in Partnership* international commissioning scheme, linking foreign-language playwrights with their British counterparts to produce international contemporary plays on the UK stage. She was Literary Manager of the Gate Theatre, London, and has worked extensively as a dramaturg.

Jonathan Meth was the Director of writernet and is now Executive Director of THEATRE IS. He has extensive experience as a freelance producer, director, script editor, reader, workshop leader and lecturer. He is also an Associate Tutor at Goldsmith's College, London, and an external assessor for the University of East Anglia's MA in scriptwriting.

Teresa Murjas teaches Theatre Practice and Polish Theatre and Film at the University of Reading (UK). She also conducts research through practice, focusing mainly on fin-de-siècle European theatre, and her performances, in new translations, have been staged at a variety of venues in the UK and abroad.

After studies in translation, anthropology and Canadian studies at the University of Alberta (Canada), **Nicole Nolette** is now pursuing a Ph.D. in French language and literature at McGill University in Montreal, Canada. Her research interests include translation studies, multilingual and sociolectal theatre, and French-Canadian literatures.

Manuela Perteghella is currently Senior Lecturer in Translation at London Metropolitan University (UK). She has published on theatre translation, and on translation as creative writing. Her doctoral thesis focused on collaborative practices of translation. She has worked for the Gate Theatre, London, and for various theatre companies, and also acted as a script reader for the Royal National Theatre Studio, London. She is co-host and co-founder of the STRAP (Stage Translation Research Adaptation Practice) discussion list.

Siobhán Rocks graduated in Theatre and English in 1984, and has over twenty years' professional performance experience particularly in the disciplines of physical theatre, Clown and Bouffon. She also has ten years' experience as a British Sign Language interpreter, and in 2006 gained an MA in Interpreting Studies from the University of Leeds (UK). She is currently a Ph.D. research student in the Centre for Translation Studies at the University of Leeds, developing a multimedia annotation tool for the analysis of British Sign Language interpreted theatre. Siobhán is a regular sign language interpreter at Manchester's Royal Exchange Theatre and Contact Theatre, and with stand-up comedian Laurence Clark, and an actor/interpreter with touring theatre company Fittings Multimedia Arts. She is also engaged in the development and implementation of training for British Sign Language (BSL) interpreters working in theatre, and collaborating with Alex McDonald in challenging accepted notions and developing new ways of working with BSL in theatre and film.

Margaret Rose teaches Theatre Studies and British Literature at Milan State University (Italy). Academic publications are mostly in the area of nineteenth- and twentieth-century and contemporary theatre. She is a translator and a writer whose translations and stage plays have been performed in Britain and Italy.

Zoë Svendsen is a director, translator and researcher. She is currently Research Fellow in Drama and Performance (practice-based) in the English Faculty at the University of Cambridge (UK), having completed a Ph.D. on aesthetic, spatial and cultural practices at the Gate Theatre, Notting Hill, London. Zoë has worked extensively internationally as a dramaturg and artistic collaborator, and regularly translates plays from German (recently on attachment at the National Theatre Studio). Zoë is Artistic Director of Metis Arts, and is co-directing and writing the TippingPoint commissioned performance installation, *3rd Ring Out*.

Carole-Anne Upton is Professor of Drama at the University of Ulster (UK). Alongside directing and translation for performance, her research interests include modern and contemporary Irish theatre, francophone drama, postcolonial and post-conflict performance and theatre for social justice. She was a founder member of the Performance Translation Centre at the University of Hull (UK). She has published on Irish and Northern Irish, African, and Caribbean theatre, and edited the volume *Moving Target: Theatre Translation and Cultural Relocation* (2000). She is founding editor *of Performing Ethos: An International Journal of Ethics in Theatre and Performance.*

Alain J. E. Wolf teaches Translation Studies and French language at the University of East Anglia (UK). He has published in the areas of pragmatics, translation studies and film adaptation. He has a current research interest in the relationship between interfaith dialogue and intercultural communication.

Introduction

Cristina Marinetti, Manuela Perteghella, Roger Baines

Within the last decade, translation for the stage has increasingly been shaping up as a significant area of research in the English-speaking world.[1] Variously known as drama translation (Aaltonen 2000), theatre translation (Aaltonen 2000, Zatlin 2005), performance translation (Hale and Upton 2000) and translation for the theatre (Bassnett 2000), academic interest in translation in the context of drama and theatre practice has been flourishing well beyond translation studies, the traditional home of research on translation. A new working group on Translation, Adaptation and Dramaturgy has been established within the International Federation for Theatre Research, while special issues of journals on both sides of the Atlantic have looked at translation as a paradigm for the exploration of blurred boundaries between national dramatic traditions, theatre histories and the set roles of theatre makers.[2]

Staging and Performing Translation: Text and Theatre Practice locates itself within this cross-disciplinary discourse as it draws on perspectives ranging from pragmatics to deaf studies, sociology of translation and performance theory, and it does so with a specific agenda: to explore the territory that exists between theory and practice. Featuring contributions by academics from theatre and translation studies, as well as translators, directors, actors, dramaturges and literary managers, this collection attempts to delineate a new space for the discussion of translation in the theatre that is international, critical and scholarly while rooted in experience and understanding of theatre practice.

While other works have sought to bring together scholars and theatre practitioners on the subject of translation (Johnston 1996, Hale and Upton 2000, and Coelsch-Foisner and Klein 2004), this volume provides radically new perspectives and moves forward from past studies as it attempts to explore and theorize the relationship between written text and performance starting from actual creative practice. It focuses on translation as an empirical process. Indeed, many of the essays are reflections of actual practice – accounts of first-hand experiences of translation and staging (Murjas, Rose and Marinetti, Benshalom) – while others explore experimental methods where translators use rehearsal techniques (in both source and target culture/language) to give space to the performative dimension of dramatic language (Baines and Dalmasso, Finburgh). Part III, 'In Conversation with Practitioners', is entirely dedicated to perspectives on practice. It includes an interview with the award-winning translator Christopher Hampton, and gives voice to the professionals who commission, evaluate and generally contribute to the shaping of translation policies and to the circulation of foreign drama (Bradley; and Meth, Mendelsohn and Svendsen).

It is indeed this very practice underpinning translation for the stage – complex, multifaceted, diverse, cultural and often personal – that translation scholars have tended to shy away from (Hale and Upton 2000, p.12), preferring to focus on how translated plays function as cultural products (Aaltonen 2000, Anderman 2005). At a time when translation studies is questioning its objects of study and its disciplinary boundaries (Bachman-Medick 2009) we take a leaf out of the theatre studies' book and look at creative practice as a model for self-reflexivity (see Lehmann, 2006, p.17). The contributions in *Staging and Performing Translation* bring practice back to the centre of discussion on translation, as they look at the complex web of collaborative processes involved in the translation, production and staging of translated plays.

The focus on practice and collaboration is not only a deliberate methodological choice of editors and contributors but is also the result of a larger vision: to create productive opportunities for dialogue and collaboration between scholars and professionals on the process of translation.[3] The fostering of such dialogue not only has offered opportunities for a deeper and better understanding of practice in the theatre, and of critical reflection on practice, but it is also of equal importance for ongoing debates in the disciplines of translation studies, theatre studies, performing arts and the theatre industry.[4] This perspective goes hand-in-hand with current methodological developments in the humanities which have called for practice-led research to encourage the integration of professional experiences and input into academic debates (Spellmeyer 2003, as quoted by Johnston in this volume).

The contributions in this volume offer a collective, multivoiced discussion of the nature of theatre translation as complex creative collaboration by exploring processes of writing, producing and commissioning translations and adaptations through a number of cultural and professional perspectives. The main themes – the relationship between text and performance and the interface of the translator and the theatre practitioner – are expounded through carefully grouped essays which explore these relations in the light of theoretical, pedagogical and practical applications complemented by linguistic and performance examples in several verbal and theatrical languages.

In this sense, this publication crosses boundaries between translation and theatre studies, and textual and performance methodologies, suggesting a need to rethink fixed models and procedures resulting from an adherence to disciplinary orthodoxies and encouraging new possibilities for future cross-disciplinary research, study and training. In so doing, it encourages us to understand translation for the stage not as a 'poor relation' of translation studies (Lefevere, 1980, p.178) or as a new interest of theatre research but as a fruitful and innovative meeting ground of disciplines.

* * *

The book is organized in four thematic sections. The first focuses on experiential and theoretical accounts of staging translations, the second on the practicalities of staging and producing translations, including the exploration of workshops, of the rehearsal space, and the visibility of collaboration – which highlights how stage translation is essentially a communal project. The third provides the perspectives of theatre practitioners, including theatre policies on commissioning and staging translations, while the fourth addresses issues of politics and ethics with regard to conflict, ideology, and gender and identity.

The contributions in Part I, 'Explorations and Experiments in Theory and Practice', all, in their different ways, undertake either to challenge a range of assumptions about theatre translation or to lay down new theoretical and practical approaches within specific contexts. David Johnston exploits his experience as academic and successful stage translator to help situate the book's central concern with creating an interdisciplinary space for the interface of translation and theatre practice both within the fractured context of humanities research and in critical conversation with relevant literature in translation studies. Most importantly, he suggests, without ever making claims to theorization, a series of metaphors and metonymies (theatre translation as a 'diaspora of form and meaning', the 'convincing' roles of the translator for performance, 'the translator as actor' – both visible and invisible in his/her relationship with the text) that offer a new language for thinking in theatre translation while mapping out a new space for it as a paradigm of reflective practice. Carole-Anne Upton's essay seeks to re-map the relationship between director, translator and performers from her experience of translating and directing a production of Maurice Maeterlinck's *Les Aveugles* in Northern Ireland, based on the contention that theatre translation may usefully be regarded as a function of *mise en scène*. She proposes a new theoretical division of the stage translation process which schematizes the foregrounding of theatrical discourse over dramatic action and weaves together the director's treatment of the text with its translation so that the two can both be seen as complementary aspects of the treatment of a play text preparatory to performance. Upton's interest in the rhythm and in the music of *Les Aveugles* is echoed in the next contribution from Roger Baines and Fred Dalmasso, the experimental nature of which lies in its focus on the translation of performance and the use of rhythm and sonority as the key that bridges performances. A close analysis of the structure and rhythm of the original text of Koltès' *Dans la solitude des champs de coton* is used as a basis for a translation method which, via the performances of actor and jazz poet David 'Stickman' Higgins, produces a text which has rhythm at its core. Both Upton, and Baines and Dalmasso thus propose new ways of engaging with the process of stage translation. Upton's model of the stage translation process is applied to a familiar type of translation process, while Baines and Dalmasso propose a more innovative translation method. Siobhán Rocks continues this tone of innovation by proposing a theory of stage interpreting into British Sign Language (BSL), thus opening up the investigation of yet another channel of mediation. It is notable how little has been written on this topic and that no sophisticated theory of sign language stage interpreting previously existed, while the practical conditions in which sign language interpreters work are revealing. This piece, which is illustrated by a series of illuminating examples from the author's extensive experience of stage sign language interpreting, sets out a range of principles for successful BSL stage translation and is likely to prove an extremely useful model for the preparation of texts for BSL interpreted performances for practitioners and theorists alike.

Alain Wolf's essay uses linguistic theories of conversational implicature and inference which have been explored in some detail within translation studies but not yet systematically applied to stage translation. These theories are used as a prism through which to analyse the translation, preparation and performance of a production of Jean Anouilh's *Antigone*. Wolf's essay functions as a hinge between the theoretical and the experiential, bridging the scholarly and the practical, preparing the reader for the creative practice explored in the second part of the volume, which focuses on accounts of 'translation in production'.

Part II, 'Practical Perspectives on Translation, Adapting and Staging', contains four essays which each take a different but distinctly practical approach to accounting for specific translated stage productions. Megan Evans focuses on the problematic of transferring the conventionalized performance forms of Chinese *xiqu* plays into English with specific attention paid to workshop practice and the synthesis of song, speech, dance-acting, combat, acrobatics and complex musical structures. Anthony Meech then provides a detailed dramaturg's perspective on the complete process of the adaptation of Brecht's *Threepenny Opera* for the National Theatre's 2002 production which incorporates the genesis of the original text, its previous productions on the British stage including productions at the National Theatre, and an analysis of the translation/adaptation on the level of text, but also on the level of design and music. Unfortunately, in this particular instance, the Brecht Estate refused to grant Meech permission to quote from his own and Jeremy Sams' translation of Brecht's play. This posed a series of problems for the author and the editors, not least how to overcome the difficulty of discussing something to which the reader cannot have access. At the same time, we felt very strongly about what we perceived to be the negation of the translators' intellectual property. We eventually decided to use what can, ironically, be considered as a species of Brechtian *Verfremdungseffekt* by leaving a blank space where the translation should have been (with paraphrased endnotes). This becomes, in our opinion, a powerful and 'performative' example of the invisibility of the translator in the field of theatre.

The following essay, Margaret Rose and Cristina Marinetti's 'The translator as cultural promoter, or how Renato Gabrielli's *Qualcosa Trilla* went on the road as *Mobile Thriller*' provides an account of the translation/adaptation, rehearsal, and performance of Gabrielli's play at the Edinburgh Fringe Festival and on tour in the UK. Here the traditional idea of the translator as mediator between cultures is taken up and expanded by showing how in the complex contexts of actual practice, translators can take on a role which extends far beyond the delivery of the text in the target language and involves presence and input in the creative processes of constructing a site-specific performance as well as acting as cultural promoter of the source culture by raising funds and facilitating contacts with partners and venues.

The remaining contribution in this section, Louise Ladouceur's and Nicole Nolette's essay '*Cow-boy poétré*: a Bilingual Performance for a Unilingual Audience', is an interesting account of the political and aesthetic complexities of translating and producing bilingual plays in Canada. The focus is on the peculiar case of Western Canada, where recent theatrical works have started displaying vernacular French-English code-switching. The authors argue how new modes of translation, particularly performance-based translation devices, are needed to both preserve the language duality of the plays and to reach an audience with different language skills.

Part III, 'In Conversation with Practitioners', opens with an interview with the renowned stage translator and adaptor (also playwright, screenwriter, director and producer) Christopher Hampton. Hampton discusses in detail what translation, including adapting 'literals', means for him, and the importance of affinity with a playwright (in his case, Ibsen, Molière, Laclos, von Horvath, and Reza). This is followed by Jack Bradley's 'Not Lost in Translation' which provides a snapshot of the state of the British stage in relation to the importation of foreign texts; the former Literary Manager of the National Theatre (1995–2007) details the considerations involved in getting non-English language texts produced. Finally, in a roundtable discussion, Jonathan Meth (former Director of writernet), Katherine Mendelsohn (Literary Manager at the Traverse Theatre, Edinburgh) and Zoë Svendsen (Artistic Director of Metis Art, and researcher) discuss their experiences of developing networks of translators and producers (The Fence, writernet and Janus), stage translation policies at the Traverse Theatre in Edinburgh, and the Gate Theatre in London. What emerges from this insightful, and often passionate, conversation is an account of the diversity of theatre translation practices and policies in the UK and in continental Europe. The discussion opens up ethical and professional implications for stage translators, as well as for practitioners commissioning and producing drama in translation.

The final section, 'Politics, Ethics, and Stage Translation', brings together three essays from Yotam Benshalom, Clare Finburgh and Teresa Murjas which, in different ways, engage with political and ethical questions. Benshalom analyses a series of ethical, ideological and political issues related to the staging of the Palestinian Riad Masarwy's *The Impotents* in his own Hebrew translation, focusing on how political over-sensitivity can damage a translated play text. The author provides a personal account of a partnership and subsequent collaboration between playwright, translator and commissioners, as well as discussing the reception of the play by a mixed audience at the Acco Festival. Whilst acknowledging that when working within contexts of conflict there is a tendency to impose a political significance on texts, he argues that this would ultimately damage the message of the text, and promotes instead a different view. While one can argue that such contexts of conflict, especially violent ones, leave little room for 'neutrality', Benshalom's essay nevertheless opens up a timely and important debate on the ethical role of translator.

Finburgh uses her work translating, in close collaboration with the playwright, Noëlle Renaude's heavily culture-bound *Par les routes* to emphasize the cultural imperialism of the UK's resistance to French playwriting in which reality lies *in* language, reality *is* language. She argues that the loyalties of a stage translation can rest less with the theatrical impact of the original, and more with the conventions governing the UK theatre establishment within which terms such as 'performability', 'breathability' and 'speakability' become ideologically inflected in ways that serve to exclude experimental theatre such as that currently produced in France. Murjas is a director-translator and her piece deals with hybrid identities as it is written from her perspective as the daughter of post-war Polish 'émigrés' to the UK and places the translation and staging process within narratives of emigration, deportation and exile and post-WWII processes of Polish/British identity formation.

As previously affirmed, one common thread that binds all the contributions together is their empirical and exploratory approach to translation in the theatre whereby translation as a practice, as well as a concept, is analysed, discussed, 'performed', through a diverse array of new, often embodied, theories. The practice-rooted, performance-based methodology of most essays embraces a new way of researching translation for the stage, opening up new avenues of cross-disciplinary research and collaboration while suggesting that more work on the interface between translation and performance practice needs to be done. Valuable avenues for further investigation that seek to provide an interface between theory and practice could include: more practice-based work on the actors' role in shaping the translated performance texts; more research on non-dramatic translation, translators working with companies devising work without scripts; together with a broader questioning of the nature and function of translation in an increasingly global and multicultural theatre market.

Notes

1. In continental Europe interest in theatre translation has existed since the 1970s, especially in the work of theatre semioticians (Patrice Pavis, Erika Fisher-Lichte, Alessandro Serpieri), see Bassnett (1980).
2. Here we refer to the *Journal of Romance Studies* (2008), an upcoming issue of *Comparative Drama* (2010) and *Theatre Journal* (2007) respectively.
3. Prominent among these are initiatives such as the JISC-hosted STRAP (Stage Translation Research Adaptation Practice) network and discussion list (https://www.jiscmail.ac.uk/cgi-bin/webadmin?A0=STRAP), the above mentioned 'Translation, Adaptation and Dramaturgy working group' within IFTR, a symposium on Theatre Translation at the University of Milan Statale in 2008 (http://www.unimi.it/indice_analitico/28807.htm), as well as the large AHRC-funded project 'Out of the Wings: Spanish and Spanish American Theatre in English Translation' (http://www.outofthewings.org/index.html).
4. Interest in the interface of academic and professional perspectives continues to be considerable. In 2010 in the UK alone there will be a 'Graduate Colloquium on Theatre Translation' at Queen Mary, University of London, a one-day event on 'Translating Theatre. Migrating Text' at the University of Warwick, and a 'Symposium on East European Dramaturgy, Adaptation and Translation' at the University of East London.

References

Aaltonen, S. (2000) *Time-Sharing on Stage: Drama Translation in Theatre and Society* (Clevedon: Multilingual Matters)

Anderman, G. (2005) *Europe on Stage. Translation and Theatre* (London: Oberon Books)

Bachmann-Medick, D. (2009) 'Introduction: the Translational Turn', *Translation Studies* (2: 1), pp.2–16

Bassnett, S. (1980) 'An Introduction to Theatre Semiotics', *Theatre Quarterly* (10: 38), pp.46–55

Bassnett, S. (2000) 'Theatre and Opera', in P. France (ed.) *The Oxford Guide to Literature in English Translation* (Oxford: Oxford University Press), pp.96–103

Coelsch-Foisner, S. and Klein, H. (eds.) (2004) *Drama Translation and Theatre Practice* (Bern: Peter Lang)

Hale, T. and Upton, C. (2000) 'Introduction', in: C. Upton (ed.) (2000) *Moving Target. Theatre Translation and Cultural Relocation* (Manchester: St. Jerome Publishing)

Johnston, D. (ed.) (1996) *Stages of Translation: Essays and Interviews on Translating for the Stage* (Bath: Absolute Classics)

Lefevere, A. (1980) 'Translating Literature/Translated Literature: The State of the Art', in O. Zuber (ed.) *The Languages of the Theatre: Problems in the Translation and Transposition of Drama* (London: Pergamon Press), pp.153–61

Lehman, H.-T. (2006) *Postdramatic Theatre* (London: Routledge)

Spellmeyer, K. (2003) *Arts of Living: Reinventing the Humanities for the Twenty-First Century* (New York: Albany State University Press)

Zatlin, P. (2005) *Theatrical Translation and Film Adaptation: A Practitioner's View* (Clevedon: Multilingual Matters)

Part I
Explorations and Experiments in Theory and Practice

1
Metaphor and Metonymy: the Translator-Practitioner's Visibility

David Johnston

> I dedicate to speech, to pomp and show,
> This playhouse re-erected for the players.
> I set my saw and chisel in the wood
> To joint and panel solid metaphors;
> The walls a circle, the stage under a hood –
> Here all the world's an act, a word, an echo.
> (Seamus Heaney, from
> 'Peter Street at Bankside')

Reflective practice

That the discussion that follows relies on practice as its touchstone may seem reassuringly concrete to some. But it is not a position devoid of its own difficulties. The discourse of practice is frequently a literal one, its sphere of reference often anecdotal, so that practice-based writing may be easily dismissed as lacking any real theoretical interest or philosophical engagement. This leads in some quarters to an over-determined view of the practitioner, where the virtue of the concrete is dismissed as the merely situational. To some extent, of course, this reflects the core critic–creative divide that, spuriously or not, still cleaves many departments of literature and theatre within the institution. Terms like 'performance' and 'theatricality' are brandished and disavowed, press-ganged into service both as the embodiment of an authentic, but fleeting and necessarily subjective, communication, and equally, as a form of deceptive synchronism that detracts from more objective and universalizing judgments. In the words of Shannon Jackson (2004, p.123), theatre itself becomes 'an index of anti-disciplinarity', while institutional practices and successive theoretical retrenchments, for the moment at least, rest more happily on paradigms rooted in the disciplinary.

This, of course, is a hugely telescoped account of a binary that has been successively and repeatedly breached and re-asserted by scholars and practitioners, modernist and post-modernist alike. But although the debate has evolved, the truth is that its core antagonism still lurks unreconstructed at the heart of many critical and practitioner responses. This in turn derives from and contributes to one of the fundamental, if generally unarticulated, principles of humanities scholarship – that is, of the perceived superiority of decontexualized textual critical analysis over creative engagement with texts. It is precisely the deracinated nature of much humanities criticism, its notable aloofness from the apparently levelling tendencies of the modern world, that pragmatic thinkers like Stanley Fish and Richard Rorty confront. Kurt Spellmeyer (2003, p.7) takes the argument one stage further when he notes: 'Our direct involvement in the making of culture – this is what the old humanities have failed to achieve and what the new humanities must undertake if they are to have any future at all.' Surely, in this context, our growing sense of crisis in the humanities cannot be dismissed as solely one of funding; it needs also to be recognized as one that arises from the entrenched autotelism of the activities that configure its mainstream disciplines.

There is of course an unacknowledged aristocracy here; presiding over all of this is a discernible resolve within the institution not simply to maintain the clearly differentiated parameters of thinking and making, but crucially to ensure that thinking is much more generously empowered.[1] Translation, however, offers a resolution to this entrenched dichotomy. As a reflective mode, translation concerns itself with the methods, meanings and movements of linguistic and cultural encounter, while as re-creative practice it engages with texts, perhaps in the most complete way possible. Now, of course, this could in turn be taken as pointing to divisions between the theorists of translation studies and practising translators. But such divisions, no matter how they are perceived, are more notional than real. Indeed, rather than being rooted in differences of hierarchy, they tend to reflect two stages of a process, two sides of the same coin. At the heart of translation, of every act or event that is generated by a translator, there is a double consciousness, a decentredness or lack of fixity that prompts, for example, Paul Ricouer to talk about the special 'aterritoriality' in which the act of translating takes place (see Kelly, 2007, p.7). Translation, of course, already straddles the apparently competing imperatives of explaining the other to the self while, at the same time, protecting the other from assimilation by the self, an enablement of two-way traffic that is implicit in translation's self-selected metaphors of bridges, doorways, portals and windows, a looking into and a looking outwards from the heart of the cultural matrix. In the specific context of literary, and of course theatre, translation, this Borgesian doubleness, the concurrent inhabitation of the here and there, of the then and now, extends into a simultaneity of thinking and doing. It is here that theatre translation establishes its claim to be considered a paradigm of reflective practice *par excellence*, because it is at this point that critical reflection and cultural praxis coincide most clearly. It is here that any established truth about a text begins to deliver multiple and frequently competing new versions of itself, a diaspora of both form and meaning that is the lifeblood of theatre, perhaps even one of the motor forces of culture itself.

To talk about translator doubleness is therefore more than a nod in the direction of the dyadic unity of translation as theory and practice. It also suggests the dual perspective inherent in how the translator approaches text, the aterritoriality of a reader-practitioner who simultaneously exists within the world of the text whilst plundering it for its potential resonances within the contours of a new target language and culture. There is a parallel here with the theatre practitioner who brings a similar kind of attention to the act of reading, one that constantly predicates movement from the logocentric to the potentially enacted, the text as written script and as encrypted performance. Of course, it has long been accepted that to translate plays is to write for performance. But even this acceptance is frequently not unproblematic in terms of its most basic assumptions. In her recently published book on theatrical translation, for example, Phyllis Zatlin (2005, p.vii) notes, by way of exemplar, that 'Marion Peter Holt, the foremost translator of contemporary Spanish theatre in the United States, affirms that performability has been the prime aim of every play he has translated.' But the central position of the book remains that of defending translators from the charge that pandering to the demands of performance will inevitably see them typecast as Iago (presumably to the author's Othello). Zatlin (2005, p.1) concludes, with defensive simplicity, that 'to achieve speakable dialogue, theatrical translators can and do adapt'.

Practice and the performable

The problem with what Zatlin has written, even taking it as a merely provisional conclusion, is that it reinforces the hierarchy of original and re-created texts – in Benjamin's terms, between aura and reproduction. Reproductive practice, it seems to suggest, necessarily compromises the integrity of the original: inevitably, it implies damage. Clifford Landers is cited in support (Zatlin, 2005, p.1): 'Even style, which is by no means unimportant in dramatic translation, sometimes must yield to the reality that actors have to be able to deliver the lines in a convincing and natural manner.' So yielding to reality is to bow to the inevitable, namely that theatre must stoop to conquer. Once again the emphasis is on the speakable. But now there is an added confusion: the 'convincing', properly considered, is the property of that domain of performance that belongs to both translator and actor. While the 'natural' may well be discernible if we filter written dialogue through our own template for naturally occurring speech, the 'convincing' is a more subtle contrivance between the natural, which persuades as an utterance in terms of its linguistic and contextual truthfulness, and the stylistic, which is credible in terms of its artistic integrity, its location (or calculated dislocation) within the voice, or style, of the play. The 'convincing' in theatre is that which turns the 'natural' into something memorable.

This leads us towards the debate on performability, into which Zatlin and Landers, with their implicit characterization of performance-oriented translation as inflicting an albeit forgivable deviation from the textual norm, have wandered almost unwittingly. As Mary Snell-Hornby (2006, p.86) notes, traditional translation theory makes an easy distinction between what she calls the 'faithful' and the 'performable' methods of theatre translation – that is difficult to argue with, although in the raggedy world of translation practice, probably every translator finds performability at times within the literal. So that even here the distinctions are blurred. Susan Bassnett, however, aims to be much more clear cut. One of her principal arguments against performability is that it cannot be conceptualized or quantified within a model. She writes (1998, p.95) that 'it seems to me a term that has no credibility, because it is resistant to any form of definition'. She further elaborates that 'attempts to define the performability inherent in a text never go further than generalized discussion about the need for fluent speech rhythms in the target text'. This is apparently echoed by something Michael Frayn (1991, p.355) writes in his introduction to his Chekhov translations: 'Translating a play is rather like writing one. The first principle, surely, is that each line should be what that particular character would have said at that particular moment if he had been a native English-speaker.' This of course would produce a version aloof from the shaping movements and perceptions of any original, in other words a wholly domesticating strategy. Compare this with Brian Friel's intentions in translating Chekhov, as a way of imagining new alternatives in inbred claustrophobic Ireland, 'of snagging the romantic ideal we call Kathleen'. So that when Chekhov is performed in Ireland, it is as a writer who is a familiar other. In the words of the actress Susan Fitzgerald (Pine, 2006, p.108): 'it's just like someone looking out of the window, but it breaks your heart'. In England, in Frayn's translations, Chekhov is performed in terms of very English silences, so that the experience is more akin to looking out of the window, but seeing only your reflection framed there.

So, while it may be difficult to theorize performability in terms of a writing paradigm or model, to present the pursuit of the performable as simply something the practitioner inevitably does, an unquantifiable element of their craft, is to essentialize practice. It is further to mystify the creative response that translation brings to text. What is certainly clear from the Frayn/Friel comparison is that the way in which a practitioner conceives of the performable may well have important implications for the extent to which any particular play in translation can provide opportunities for its spectators to engage creatively and meaningfully with a culture other than their own. In one way, of course, Landers is right to insist upon the convincing as a defining element of the performable. Performance has to be bought into. But the quality of being convincing is one that is contingent more upon the particular terms of engagement that a play or theatre event proposes to its audience than on any lingering sense of fluidity or naturalness.

This is clearly true of any original play. But when it comes to the conceiving and writing of the play in translation, there are additional opportunities and pitfalls. At the most general level, how we consider and think about translation for performance has more to do with theatre practice than it has with translation theory *stictu sensu*. That is not to deny the clear and potentially very fruitful relationship between theory and practice, but to subordinate performance-led practice to a hierarchy of theory – which is what Zatlin effectively does, albeit represented here by a paradigm founded implicitly on an uncomplicated notion of equivalence – that effectively serves to decontextualize practice from the imperatives that drive it. A translated play is already a work in transition, so that its responsiveness to the contingency of the performance environment, a key factor in shaping the reception of any play, is correspondingly more open, more malleable. It is precisely in this openness, this malleability that opportunity and pitfall frequently coincide.

Practice and its contexts

Translation for performance as a writing practice and theatre as a collaborative making-practice are about how we place contexts around actions; no other form demonstrates so completely the effects of contingency. By highlighting the constantly changing nature of forces in play at moments that are in themselves various and ever-shifting, theatre and translation ensure that these contexts – we could also call them frames – cannot be categorized or seen in terms of any given model or be subject to any essentialist understanding. Translation and theatre encourage us to relativize not just the apparent truths of any given situation, but also the frameworks in which those truths are dramatized. Crucially, therefore, translation and theatre locate its practitioners as non-centred points in an ever-fluctuating network of activity. In other words, translation practice is a form of cultural agency that, like all cultural production, is dispersed across a range of economic, linguistic, historical and intellectual models, among others. This breadth of impact, this need to harness multiple practices and effects, may, of course, be one of the reasons why translation necessarily deploys so many metaphors in its own self-reflective discourse. But what is certainly true is that theatre translation is a mobile practice in which nothing is fixed, everything asserts itself to everything else in and through constantly changing frameworks. Translation and theatre narrativize the ways in which space and time are contested qualities, so that an aliveness to translation and theatre alike brings home to us how the meanings encased in any text, of any time, shift according to the viewing frame, our sense of context.[2]

If we accept that translation is indeed such a mobile practice, one that establishes ever-changing channels of communication between the context of the writer, and those of the translator and his or her audiences, then to speak of the dialogical position of the translator between author and receiver, or between source and target texts, implies a fixity that the translator may occupy only in theory. In reality, translators are embodied subjects who extend each individual text, framed by their own context, in and through a re-creative practice that is rooted in empathy towards, and mimesis of, an object that is in itself evolving on its journey through time and space. Put in those terms, this may seem to be a statement of practitioner invisibility. But while translator-practitioners are indeed interpretive agents, the tools of both their interpretation and their agency are heavily contextualized, contingent upon their condition as embodied subjects.

The translator-practitioner, in terms of his or her relation to the text, may in this way be conceived of as an actor who performs in terms of the imperatives of the text, and who, by extending this performance to new audiences, produces work in which he or she is both visible and invisible – simultaneously subsumed into the text (actor as character, translator as reader) and an active agent of its re-creation (actor as performer, translator as theatre-writer). To put it in the most direct terms possible, the same translator will approach a Molière to be performed in Belfast very differently from a García Lorca to be performed in Coventry. But in both cases the creative lynchpin of the writing will be the awareness of audience and of impact.

A particular quality of much recent theatre theory has been the conception of space and place as active participants in performance. Both space and place have a life of their own. The environment in which performance takes place is inevitably saturated with meanings, some or many of which are capable of deep connection with the itinerant text. How the translator constructs, deconstructs or re-constructs – or perhaps chooses to ignore – those connections is achieved through a whole series of different writing strategies. But to work with the contingent framework suggested by those connections is one of the most potentially fruitful engagements of the translator, and may produce one of the potentially most rewarding kinds of night out in the theatre for the spectator. Translation for performance is well placed to restore energy to time and space, to move beyond the aporia at the heart of how we imagine time as a bifurcation between past and present, and space as a site of single identity occupation. A translated play traffics in – or at least has the potential to traffic in – the materiality of both time and space. To ignore that potential – which is the writing choice of some translators – is to work from a translation strategy that is either aloof from or subject to the complex configurations and the multi-layered histories of their own cultures. The 'aloof from' and the 'subject to' have profound implications for the shaping strategies for the writing of translations. On the surface, they can be seen as the writing tactics suggested by the extremes of domesticating and foreignizing, but once again, in practice, they have as much to do with how the translator situates him or herself both in relation to the cultural work of the text in question and to the locus of its reception.

This is the cultural analysis which the translator brings to the play to be translated, in part critical practice and in part re-creative strategy. The term 'cultural analysis', as it is used here, is borrowed from the idea of 'cultural memory in the present' that Mieke Bal develops in the context of historiography. Bal (1999, p.1) notes:

> Cultural analysis as a critical practice is different from what is commonly understood as 'history'. It is based on a keen awareness of the critic's situatedness in the present, the social and cultural present from which we look, and look back, at the objects that are always already of the past, objects that we take to define our present culture.

In terms of translation practice, the central concepts here are as equally applicable to geography as they are to history. Just as cultural analysis, in Bal's conception, seeks to understand the past as a living part of the present so, through the work of the translator, it can also create ways by which the assumptions and practices of other cultures are infused into the assumptions and practices of an audience situated in the here and now of performance.

Practice and metaphor

The stage translator, like all of the other practitioners who collaborate in the making of theatre, is centrally concerned with constructing performance. In that sense, performability, as the quality that ensures the play's success in stimulating and sustaining the authorized game of make-believe that is theatre (Walton, 1990) is as much a default concern of stage translation as expressive writing is of the novelist. Performability is an implicit part of any re-creative strategy, and, as such, is certainly open to analysis. Indeed, that is one of the central purposes of this chapter. But whether to debate its validity as a concept is fruitful or simply nugatory is another matter.

In the final analysis, whether translators consider performability to be an active concern of the translation process, or the proper preserve of director and actors, will depend markedly upon the extent to which they view themselves as an active collaborator within the dynamic process of staging a play. And what these active collaborators may achieve, at their best, is to ensure that something from somewhere or sometime else is not only meaningful in the experience of a spectator who exists in the here and now of performance, but that that artefact also continues to belong to itself. Translation, and especially translation for the theatre, is a process that in this way engineers two-way movement – a traffic between the narratives, concepts and structures of life embodied in foreign texts, and the affective and cognitive environment of the spectator.

By working as a writing conduit between two conditions of situatedness, that of the text in its time and place, and that of the spectator in the present moment of performance, the translator provides passageways of thought and feeling, sometimes on the surface, sometimes profound, between the here and the elsewhere, the now and the elsewhen. Put another way, there is often no need for the translator to have to elect between the domesticating and foreignizing strategies that are still seen as enshrining commercial and purist approaches respectively, because in practice he or she can choose to employ both strategies simultaneously. The translator's shifting gaze allows the text to be simultaneously of then and there, encased in cultural difference, but also belonging to the shifting here and now of our spectator. In other words, translation is not a filter between past and present, for the cultural other and the located self; it is potentially a prism that releases, that fires off in different directions a series of intercultural and intertemporal moments that challenge and enrich spectator reception and experience. In that way, the translated play comes into its own; by not masquerading as a piece of English theatre manqué, by asserting its newness and originality in every major production, the translated play begins to stake a claim for its own special place in the theatre today. What translation can do most powerfully in this regard is to promote hybridity, a hybrid text that simultaneously moves between and across different histories and geographies, locating and uprooting the historical and cultural imagination of the spectator in a way that seeks to overcome the twin separations implied by aloofness and subjection.

This is the essence of the translator-practitioner's creative struggle – a term, of course, that is no more capable of being elevated to the status of paradigm than performability itself, because both, though intimately linked, are rooted in the specifics and contingencies of individual plays and their relationships to individual performance environments. Whatever solution the translator-practitioner may bring to the conundrum of extending the foreign play to another theatre system, while at the same time enabling it to speak vividly of its own different context, will invariably be judged as a foundation element of the production, either enriching or hindering spectators' engagement with the play in question. The consequence of that is that the translator-practitioner's visibility now enters the purview of critical review, where it arguably always belonged. But the translator's right to search for such solutions lies in the status of translation as a writing practice that eschews fixity, or locatedness, and as an ethical regime that is anxious to preserve the claims of alterity. So that if and when the translator rejects metonymical forms of reproduction as being rooted in unsustainable notions of commensurability, the new text he or she produces will function as a metaphor bridging two contexts, a perception of the similar between two dissimilars.[3] It is on the generation of the metaphorical relationship between source and target texts that the real creativity of the translator is focused. Film studies scholar David S. Miall (1987, p.82) notes that 'metaphor shows on a small scale all the principal features of the thought processes that are most significant in creativity'. At the heart of good metaphor, and of our response to it, are detectable what Miall (1987, p.82) notes as the constituent elements of creative thought – 'the presence of productive anomalies, the crossing of conceptual boundaries, the transformation of a subject within a given domain, and the intuition of a new order at a moment of illumination'. These of course are no less the results of the thought transformation processes that operate within the re-creative strategies of translation. As the spectator is prompted by the contrived doubleness of the translation to engage with the metaphorical relationship that the play offers between times or places, so the degree of his or her imaginative complicity becomes increasingly active. It is the heuristic force of metaphor, its capacity to generate dramatic tension and semantic shock, that stimulates imagination and prompts understanding; surely what we are describing here also constitutes the real action of theatre? Philosopher Richard Kearney (1998, p.141) notes:

> The traditional opposition between *theoria* and *praxis* dissolves to the extent that 'imagination has a projective function which pertains to the very dynamism of action'. The metaphors, symbols, and narratives produced by imagination all provide us with 'imaginative variations' of the world, thereby offering us the freedom to conceive of the world in other ways and to undertake forms of action which might lead to its transformation. Semantic innovation can thus point towards social transformation.

Kearney's words are no less applicable to the series of dramatic movements that together vivify spectator response. The central strategic goal of this action is to allow the translated play to make its mark in the air between stage and audience. When we translate a play from somewhere or sometime else – in other words, when we translate a text, replete with alternatives and alterity – we bring into the auditorium a swirling constellation of possibilities. And of course writing or directing (and therefore translating) a play is about attempting to conjoin and coordinate this vast range of possibilities – which, taken together, constitute the gamut of the work's translatability – a quality rooted not in the notional commensurability of texts, languages or cultures, but in the metaphorical sweep of the text. Taken as a whole this vast range of possibilities represents the cultural momentum of the text, its journey through time and space, along which it acquires and absorbs different meanings and potentials for performance. And translation as an operation – or as a series of operations – of course plays a determining role in the maintenance of this momentum. There is a converse to this, which is why some plays, even successful ones, present themselves as untranslatable. In the simplest of terms, they are only what they seem to be. Compare, for example, Lorca's classic *The House of Bernarda Alba* with Martin McDonagh's *The Beauty Queen of Leenane*, plays that both centre on the savage relationship between mother and daughter(s) in a bleak and inhospitable landscape. But while Bernarda is a mother whose oppression of her daughters is recognizably shaped by and reflective of age-old patterns of political tyranny and personal cruelty, the dysfunctional relationship between Maureen and her mother Mag is little more than a caricature of childish spite. Both plays are equally concerned to portray the murderous consequences of such negative circuits of being, but the truth is that while the metaphorical sweep of *La casa de Bernarda Alba* is that of a work of art, *The Beauty Queen of Leenane* remains that of a well-crafted play.

Practice and its re-creative strategies

There are, undoubtedly, as many re-creative strategies for extending and engaging with the metaphorical sweep of a play as there are individual plays worth translating. Writing craft is, of course, crucial, because at the heart of each strategy, as it emerges from a sense of the performance potential of each individual play, are the twin objectives of re-theatricalizing the play for a new audience, whilst at the same time de-familiarizing in a meaningful way aspects of its form or frame of reference. In terms of both stage language and situation, plays tend to present a shifting balance of the familiar and the unfamiliar, each serving as an optic into the other, together drawing the spectator into a journey that is simultaneously into the self and the other. This is perhaps one of the principal goals of the re-creative strategies of a translation method that is concerned to prompt the spectator 'to blend in and out' of the different cultural locations, assumptions and practices proposed conjointly by source and target texts (Fauconnier and Turner, 2002).

I have been at pains to emphasize that solutions, devices, interventions, whatever we choose to call them, work best when they are seamlessly dramatized extensions of moments or issues dramatized already within the original. Genre, of course, has a central role here. In comedy, for example, the translator can more easily resort to anachronism to prompt the spectator to move between different historical moments. Here is an example taken from a recently commissioned translation of *The Miser*.[4] Set now 'Somewhere in Eighteenth-Century England' the play seeks to project configurations of this nascent capitalist society against the economic downturn that unleashed itself in 2009 as yet another in a series of recurrent crises of the modern banking system. Harpagon (now Harpingon) has just manhandled the innocent servant La Flèche (now Jack) from his house, suspecting him of harbouring miscreant intentions:

Harpagon: Adieu. Va-t-en à tous les diables!
La Flèche: Me voilà fort bien congédié.
Harpagon: Je te le mets sur ta conscience, au moins.
Harpagon: Voilà un pendard de valet qui m'incommode fort; et je ne me plais point à voir ce chien de boiteux-là. Certes, ce n'est pas une petite peine que de garder chez soi une grande somme d'argent; et bienheureux qui a tout son fait bien placé, et ne conserve seulement que ce qu'il faut pour sa dépense! On n'est pas peu embarrassé à inventer, dans toute une maison, une cache fidèle; car pour moi, les coffres-forts me sont suspects, et je ne veux jamais m'y fier. Je les tiens justement une franche amorce à voleurs, et c'est toujours la première chose que l'on va attaquer.

Harpingon: We could save all this palaver
If you just tell me where it is.
You'll feel better if you confess.
Jack: Where what is?
Harpingon: It. What you've stolen.
Jack: I haven't stolen anything.
Harpingon: Thank God. I got here just in time.
Then bugger off. Before you do. Out!
(He manhandles Jack out)
Vigilance saves the day once more.
But can vigilance be enough?
When your servants are all jailbirds
like that reprobate . . . you can watch,

> you can listen, you sit all night
> and you hear the floorboards creaking,
> and you know someone's on the prowl . . .
> this is the torment of the rich –
> that it'll all be taken away.
> We've a duty of care to our cash.
> You can't trust banks, and they charge you
> every time you go near your money.
> Your own money! And as for safes,
> they draw burglars like bees to money . . .
> . . . honey, I meant.
> You might as well put up a sign:
> Money in house. Please help yourself.
> It's a worry: those thousand sovereigns
> that I buried in the garden . . .

The eight-syllable line here enacts an unfamiliar rhythm of performance, speeding up actors' delivery in the process, and is designed as a key element of timing. Rather than being an evocation of the fearfulness of the miser, the play now works towards enabling the imagination of the spectator to recognize the workings and impact of capitalist greed in two moments. To achieve this, the translator needs to further the sweep of Molière's notably concise language – a concision that the writer-actor famously extended through his own larger-than-life performance style. The old adage of stage writing – 'don't tell, show' – is put into effect as the Miser's plaintive descriptive 'Certes, ce n'est pas une petite peine que de garder chez soi une grande somme d'argent; et bienheureux qui a tout son fait bien placé' is extended into a series of nervous actions that together constitute the no less plaintive 'torment of the rich'.

The translator must, of course, provide the scaffold upon which the actor constructs his or her performance. Inevitably then the translator seeks to read and subsequently to re-write the phonetics, punctuation, and kinetic patterns of the original text, make the language performable in terms of ensuring both speakability and the special dramatic marking that comes from the lingering sense of otherness. One example, taken from the most frequently performed and translated of Lorca plays, *Bodas de sangre* (*Blood Wedding*), will illustrate this. In the opening scene of the play, the Mother curses 'la navaja, la navaja [. . .] y las escopetas y las pistolas y el cuchillo más pequeño, y hasta las azadas y los bieldos de la era' as items that all represent danger in the world of men. This is translated almost literally in the first Penguin translation of the play (Luján and O'Connell, 1961, p.33): 'Knives, knives [...] And guns and pistols and the smallest little knife – and even hoes and pitchforks.' Ted Hughes (1996, p.1) has: 'The knife, the knife! [...] And guns and pistols, even the tiniest little knife, even pitchforks and mattocks.' Brendan Kennelly (1996, p.11) widens the curse to 'The knife, the knife. [...] And the curse of God on guns, machine guns, rifles, pistols ... and knives, even the smallest knife ... and scythes and pitchforks.' In his 1980 Spanish-language edition of the play, the distinguished Hispanist Herbert Ramsden (1980, p.74) notes that the farm implements Lorca mentions 'take both their basic meaning and their emotive resonances from a cultural complex different from our own' and puts forward a number of possible translations – 'drag-hoe', 'pick-axe', 'winnowing-fork', 'pitch-fork', all of which, he argues, will permit English readers (of his published edition) to process the text from within a familiar context. Leo Hickey (1998, p.50) takes an opposing view:

> a translator can attempt either to bring the ST to the reader, with all its locutionary, illocutionary and perlocutionary import, wherever the reader may be, or else take the reader, complete with any baggage of cultural or linguistic background that may be attached to such a person, into the world – the linguistic world – of the ST. And I am suggesting that perhaps in the case of these three plays [*Blood Wedding*, *Yerma*, *The House of Bernarda Alba*], the tactic of taking the reader into the ST world should be considered.

But what is going on here in terms of performance? Given the fact that the Mother's invective comes as the first moment of heightened tension in the play – previously we have had only eleven short speeches of deliberate domestic banality – these are clearly key lines, and really have to be viewed from the overarching perspective of the play – its energy, its dominant motifs, and its meanings – as a whole. Moreover, the translator of drama for performance does well to bear in mind that whatever we might consider to be the indivisible unit of dramatic construction – the individual speech or the individual exchange – it is stamped with purpose. It is a cellular unit that carries within it the shape and force of the play in its entirety. If that cellular structure of dramatic writing is ignored, there is a real risk that the play will lose coherence, on the page and on stage alike, and will be experienced in a piecemeal and de-energized way. So the emotional action of *Blood Wedding* begins, as it will end, with an image of the knife, creating a sense of violence that overhangs the play like a damoclean sword.

'Navaja', 'knife', is a continually recurring sign in Lorca's poetry, plays and drawings, taken, as are so many of his motifs and icons, from a reality that is both observed and part of a recognized cultural tradition. The word is invested here with an elemental force that is operative both within the experience of the character herself and within the collective imagination of the audience. If one were translating the force of the word into an Irish situation, then its direct equivalent would be the gun. Both knife and gun are readily intelligible correlatives for a certain type of social and historical violence, both potent agents and harbingers of a destruction whose causes are known to all. In other words, the first mention of the 'navaja', leading as it does into this list of dangerous weapons and implements, creates a moment of expectation and of recognition; the audience begins to confront the tragedy of a relentless chain of cause and effect that it recognizes as being its own trauma. It is this act of complicitous recognition that the Irish poet Brendan Kennelly seeks to re-create by broadening his references to include 'rifles' and 'machine guns'. Indeed, his sense of the parallel between the violent divisions of Irish history and this community that bays for its own blood in Lorca's play, is reinforced by the new lines with which he has his version end. In the closing lines of the play, the Mother's references to 'this blood-haunted place' and her 'dream of peace', with their overtones of the Northern Irish peace process (it was performed in 1996), bring his version full circle, and re-create a sense of the cultural utility of the Lorca original within Kennelly's own commitment as an Irish writer. Clearly, however, it would be impossible to translate 'navaja' as 'gun', and Kennelly is sensitive to the fact that one of the principal strands in the play's grammar of imagery is that of images of cutting, pinning, slicing and piercing. Moreover, while, admittedly, it may speak of a similar macho-style response to historical dislocation, the gun does not have the specifically phallic overtones of the knife, and the sexual connotations of the death of the two men in Act Three would be lost. In this particular case, Kennelly points up connections between Lorca's project, and his own, without allowing the play to be flooded with a spurious Irishness. The spectator's imagination is located precisely where it should be: not in Hickey's Andalusia or in the comfortable familiarity of Ramsden's England, but in the theatre, the liminal space between stage and auditorium, where it belongs.

Having started with the most emotionally loaded motif – the 'navaja' – the Spanish can afford to bring in less elemental items – the 'azadas' and the 'bieldos de la era'. Clearly, the issue that the translator requires to negotiate here is whether the specificity of reference should be retained in order, presumably, to mark the difference of the play's setting – and implying, in the process, that this is the sort of thing that goes on in the Spanish countryside – or whether the referents should be strengthened in order to reinforce the energy surrounding the knife. This issue, moreover, cannot be considered in isolation from the cultural play of language that gives Lorca's plays what Kennelly (1996, p.7) calls their quality of 'rhythmical and emotional revolution'. Their characteristic linguistic actions – rhythm/repetition, the use of anticipatory poetics, kinetics, kinesics, language that is simultaneously located and dislocated – all need to be considered as an informing aspect of the overall process of cultural negotiation so that the play can be understood without being normalized. Stage language does not simply mean: it *does*. Indeed, this is surely what 'performability' is all about – giving actors lines that are speakable and that, at the same time, recreate the stylistic marking and cultural significance of the original.[5] Lorca almost certainly chose 'azadas' from the bewildering array of rural cutting tools at his disposal because of its assonant relationship with the preceding 'hasta las' and, more crucially, with the word 'navaja' itself. Moreover, the falling rhythm of 'los bieldos de la era' allows the actor in question to vary the emotional stress of the phrase so that it ends on a note of apparent helplessness in the face of omnipresent destruction. In terms of sound patterns, therefore, the specificity of these items is expendable. My own solution (Johnston, 1989, p.4) emphasizes the rhythmical nature of the language: 'I hate knives [. . .] Knives, guns . . . sickles and scythes . . . ' and in the process relies on the heavily imagistic nature of Lorca's language and its commitment to the deeply-felt expression of emotion to create a profound but intelligible sense of otherness for the spectator.

Kinetics – and to a lesser extent kinesics – are central to this creation of performable rhythms – kinetics, in terms of the way in which words are matched to movement, and kinesics, in the way in which words create spaces for non-verbal communication. The opening scene of Calderón's *El pintor de su deshonra* offers a good example of the ways in which writing implies movement and gesture. Don Juan has just arrived at the house of his old friend, Don Luis, bringing important and eagerly awaited news:

Don Luis: Otra vez, don Juan, me dad
            y otras mil veces los brazos.
Don Juan: Otra y otras mil sean lazos
            de nuestra antigua amistad.
Don Luis: ¿Cómo venís?

Don Juan: Yo me siento
tan alegre, tan ufano,
tan venturoso, tan vano,
que no podrá el pensamiento
encareceros jamás
las venturas que poseo,
porque el pensamiento creo
que aún ha de quedarse atrás.
Don Luis: Mucho me huelgo de que
os haya en Nápoles ido
tan bien.
Don Juan: Más dichoso he sido
de lo que yo imaginé.

There is a beautifully struck balance here between two men, one bursting with news, the other consumed by curiosity, and the demands that is placed upon them by codes of courtesy. It is important that this does not degenerate into mere word play because it is the same balance – between the affairs of the heart and the exigencies of an other-directed society – that will shift fatally as the play develops. Here is one version (Paterson, 1991, p.17) that claims to be 'agreeable to read and to perform':

Don Luis: Once again and another thousand times so,
I welcome you with open arms, don Juan.
Don Juan: May this and a thousand more again
bind our friendship from so long ago.
Don Luis: How goes it with you, friend?
Don Juan: I feel so happy, so gratified,
so pleased with life, so deeply satisfied,
that thought will never in the end
find the means fit to express
the sheer good fortune I possess,
for even thought I find,
will linger far behind.
Don Luis: I've overjoyed that things have gone so splendidly
for you here in Naples.
Don Juan: In actual fact, my luck is greater than
I imagined it to be.[6]

The translation is already twenty-five per cent longer than the original, struggling to clinch rhymes as well as to communicate every perceived nuance of the original. Semantic overloading is, of course, a difficulty common to many translations, especially of poetry. In this case, the formal welcome and response are excessively prolonged creating a simple problem of kinesics – the scene demands that the friends embrace before the more intimately probing '¿Cómo venís?' – translated here by the less urgent 'How goes it with you, friend?' Furthermore, Calderón is a playwright who delights in the rapid build-up of dialogue – the device of constant intercutting between interlocutors is not uncommon in his theatre – and in this short excerpt there are already two examples of lines being eagerly finished by the other speaker that this version chooses to ignore.

Laws that are no laws

It would of course be impossible to catalogue all of the types of intervention that a translator may make in order to ensure that a play from a different theatre system works within the present moment of performance. But the objective of such interventions is clear. Within the very broad goal of making a play work on stage, meaning by that a play that speaks to new audiences whilst simultaneously belonging to itself, the writer who has any regard for translation as an ethical regime traffics in what Miall calls 'productive anomalies' – a term that echoes Steiner's evocative core concept of elucidative strangeness. These carefully contrived anomalies stimulate – or, at least, are designed to stimulate – the crossing of conceptual and cultural crossing boundaries, the looking outwards from the heart of the language system, of the cultural matrix. When the translated play achieves that, as sometimes it can, it creates a different quality of theatre experience whose keynote qualities are those of the telling metaphor – the transformation of the known, the illumination that comes from contact with the unknown, the unexpected. So that, for example, critics and audiences alike sense that writers like Molière, Calderón and Lorca, through their translators, bring something qualitatively different to the English-speaking stage. Writing for this degree of performance signifies that the translator is, in this sense, a writer, and at every stage of the production process must function as a writer. In the case of theatre, the creativity of the writer, or translator *qua* writer, is not limited to the prior preparation of a blueprint for performance, but instead is more consistently and certainly more wholly engaged in the interactive practice of theatre-making which, like all interactive practices, is subject to a continual process of cultural re-evaluation. Translation study that ignores the creative imperatives of writing (rather than doing) a translation runs the risk of marauding into the fruitless, indulging in what Friedrich Dürrenmatt (1976, p.61) called the 'obstinate proclamation of laws which are no laws'.

Notes

1. Edwin Gentzler (2001, p.15) is particularly swingeing in his view that I. A. Richards' work on translation reads 'as a desperate play to retain power within the institution in light of new theoretical developments'. Such power is rooted in the control of language itself, a control that in the specific context of translation, maintains itself through an intertextual paradigm of clear subservience. In contradistinction, empirical evidence suggests that, as Gentzler observes, translations tend to respond to 'laws that are unique to the mode of translation itself', and in doing so they open up new heterogeneous perspectives that breach the barriers of our linguistic and cultural homogeneity.
2. Of course it was Einstein who insisted seminally that the relativity of the frames of reference be included in the object studied, an idea that Cervantes had initially explored so playfully in Don Quijote. See also Marvin Carlson (1989, p.15): 'The physical surroundings of performance never act as a totally neutral filter or frame. They are themselves always culturally encoded and have always – sometimes blatantly, sometimes subtly – contributed to the perception of performance.'
3. This is the broad definition given to metaphor by Aristotle in the *Poetics*. Ricoeur notably counsels against seeing this as merely an association of ideas. See, in particular, Ricoeur's essay, translated by Kathleen Blarney and John B. Thompson, 'Imagination in Discourse and Action' (1994). Tellingly, Ricoeur's *La métaphore vive* has been translated into English as *The Rule of Metaphor*.
4. Unpublished version, first performed at the Belgrade Theatre, Coventry, January 2010.
5. See, for example, Juliane House (1997). Hickey (1998, p.51) insists, properly, that 'marked should be translated as marked'. However, his argument is weakened by his reduction of the issue to the simplified question of 'whether a translation should preserve the markedness of the original or recontextualize it into something unmarked in English'. There are complex issues surrounding performance reception and the hybrid nature of the translated text that need to be borne in mind here.
6. It is the book's back-cover that claims that the translation is 'agreeable to read and to perform'. The edition is bilingual.

References

Bal, M. (ed.) (1999) *The Practice of Cultural Analysis: Exposing Interdisciplinary Interpretation* (Stanford: Stanford University Press)

Bassnett, S. and Lefevere, A. (1998) *Constructing Cultures. Essays on Literary Translation* (Clevedon: Multilingual Matters)

Carlson, M. (1989) *Places of Performance. The Semiotics of Theatre Architecture* (Ithaca: Cornell University Press)
Dürrenmatt, F. (1976) *Writings on Theatre and Drama* (London: Cape)
Fauconnier, G. and Turner, M. (2002) *The Way We Think: Conceptual Blending and the Mind's Complexities* (New York: Basic Books)
Frayn, M. (1991) 'A Note on the Translation', Anton Chekhov, *Plays* (London: Methuen), pp.xi–lxix
Gentzler, E. (2001) *Contemporary Translation Theories* (Clevedon: Multilingual Matters)
Hickey, L. (1998) 'Pragmatic Comments on Translating Lorca' *Donaire* (11), pp.48–54
House, J. (1997) *Translation Quality Assessment: A Model Revisited* (Tübingen: Narr)
Hughes, T. (trans: 1996) *Blood Wedding* (London: Faber)
Jackson, Sh. (2004) *Professing Performance. Theatre in the Academy, from Philology to Performativity* (Cambridge: Cambridge University Press)
Johnston, D. (trans: 1989) *Blood Wedding* (Sevenoaks: Hodder and Stoughton)
Kearney, R. (1998) *Poetics of Imagining: Modern and Post-Modern* (New York: Fordham University Press)
Kelly, S. (2007) 'The Island That Is Nowhere or Cultural Translation – A Utopian Project?' in S. Kelly and D. Johnston (eds.) *Betwixt and Between: Place and Cultural Translation* (Newcastle: Cambridge Scholars Press), pp.2–20
Kennelly, B. (trans: 1996) *Blood Wedding* (Newcastle: Bloodaxe)
Luján, G. and O'Connell, R. (trans: 1961) *Lorca. Three Tragedies* (Harmondsworth: Penguin)
Miall, D.S. (1987) 'Metaphor and Affect: The Problem of Creative Thought', *Journal of Metaphor and Symbolic Activity* (2.2), pp.81–96
Paterson, A. K. G. (trans: 1991) *The Painter of His Dishonour* (Warminster: Aris and Phillips)
Pine, R. (2006) 'Friel's Irish Russia', in A. Roche (ed.) *The Cambridge Companion to Brian Friel* (Cambridge: Cambridge University Press), pp.104–17
Ramsden, H. (ed.) (1980) *Federico García Lorca: Bodas de sangre* (Manchester: Manchester University Press)
Ricoeur, P. (1994) 'Imagination in Discourse and Action', K. Blarney and J. B. Thompson (trans.) in G. Robinson (1994) (ed.) *Rethinking Imagination: Culture and Creativity* (London: Routledge), pp.118–36
Snell-Hornby, M. (2006) *The Turns of Translation Studies. New Paradigms or Shifting Viewpoints* (Amsterdam: John Benjamins)
Spellmeyer, K. (2003) *Arts of Living: Reinventing the Humanities for the Twenty-First Century* (New York: Albany State University Press)
Walton, K. (1990) *Mimesis as Make-Believe: On the Foundations of the Representational Arts* (Cambridge, Mass.: Harvard University Press)
Zatlin, P. (2005) *Theatre Translation and Film Adaptation. A Practitioner's View* (Clevedon: Multilingual Matters)

2
The Translator as *metteur en scène*, with Reference to *Les Aveugles* [*The Blind*] by Maurice Maeterlinck

Carole-Anne Upton

Dear Maurice,

First run-through today. It was good. A good account of the work so far, but not a spectacular break-through. I think the actors were pleased and encouraged. I felt disappointed – not with them, but with myself. Sometimes, after a run-through, the director can see the play. [. . .] Today, it was murky and I couldn't see much.

(Stafford-Clark, 1989, 159)

This isn't really a letter to Maurice. It's part of a letter originally addressed to George Farquhar. And it's not from me, it's from Max Stafford-Clark. In *Letters to George* the director, accustomed to teasing out the uncertainties of new work in the presence of the playwright, finds himself working on a production of Farquhar's 1706 play *The Recruiting Officer*, and sets out an epistolary account of his daily process of creating a performance text on the basis of a culturally remote and necessarily incomplete play text.

My project was to create a performance text on the basis of a culturally remote, necessarily incomplete, and foreign-language play text, *Les Aveugles*, or *The Blind*, a late nineteenth-century play by the Belgian symbolist playwright Maurice Maeterlinck.

My letter would have gone something like this:

Dear Maurice,

First run-through today. It was OK. A good account of the work so far, but not a spectacular break-through. I think

> *the actors were a little lost. I felt disappointed – not with them, but with myself. Sometimes, after a run-through, the translator can hear the play and the director can see it. [. . .] Today, it was murky.*
>
> *I couldn't grasp the right register for performance. Is this a quality of your text, of my translation, my staging, or my work with the actors? There's a quiet intensity in your work that dissipates with the slightest banality in the inflection of a gesture or phrase, and implodes as soon as the contrivance of rhetoric and formality becomes too conspicuous. Just time to redraft before tomorrow's rehearsal.*

Les Aveugles was published in 1890 and first performed in 1891 in a production with Lugné-Poe for the symbolist Théâtre d'Art, loosely under the direction of Paul Fort, in Paris in 1891. It is a short piece, amongst the most successful of Maeterlinck's early works, in which his symbolist concepts for a Theatre of Stasis and a Theatre of Silence (both set out in 'Le Tragique quotidien', see *le Trésor des Humbles*) are clearly developed. It is quite startling in its apparent simplicity – of structure, situation and language, which all seem to strive for a level of rarefied abstraction that defies the inescapable materiality of theatrical presentation. The notion of character is virtually absent, as is action in the traditional sense, and there is almost no plot. The narrative centres on the performance of waiting, and listening, moving swiftly into the realm of the metaphysical, in a striking prefiguration of *Waiting for Godot*. A group of twelve blind people, most of them old, have been led into a forest by their guide, a very elderly priest, on whom they are utterly dependent. In their blindness, they wait anxiously for his return to guide them back to the home,[1] not realizing that he has already died in their midst. On discovering the corpse, the blind are forced to face their hopeless destiny: lost in the dark forest without a guide, as the snow begins to fall, somewhere on an inhospitable island surrounded by cliffs and rising seas. The play ends with the sound of approaching footsteps, the sighted baby screaming, and the blind praying for mercy.

In translating and directing the piece, I wanted to explore the relationship between translation for performance and directing, based on the contention that theatre translation may usefully be regarded as a function of *mise en scène*. In the words of Antoine Vitez (cited by Meschonnic in Déprats, 1996, p.62), 'Pour moi, traduction ou mise en scène, c'est le même travail, c'est l'art du choix dans la hiérarchie des signes', [To me, translation involves the same process as directing; it's the art of choosing from amongst the hierarchy of signs[2]].

Taking as a starting point the fairly commonplace analogy between translation and directing (see for example Zuber-Skerritt 1984, Boswell 1996, Vitez 1996 (in Déprats), Upton 2000), and assuming that the two processes can indeed be regarded as analogous in semiological terms, I wanted to explore two aspects in particular:

1. The extent to which slippage between verbal and non-verbal signifying systems could be used as a translation/direction strategy.
2. The pragmatics of a process for realizing that strategy in rehearsal; interweaving the twin roles and functions of translator-director in preparing a text for performance.

Verbal and non-verbal signifying systems

The language of the original text is simple, rhythmic, and from the outset, open to connotative readings. The rhetorical device of unanswered questions, literally echoing as they are repeated in the space immediately sets up a metaphysical theatrical discourse. The metaphors of blindness, and darkness, of losing one's way, with all the spiritual, ontological and epistemological associations of those terms are well-established and recognizable, and are held in tension, through the simplicity of the dialogue, with the naïve questions of the characters in the dramatic situation. The metaphysical discourse is utterly dependent upon the persistent absence of answers to the questions, which can be perceived in performance only through silence.

Dear Maurice,

I've decided to work with the actors on a process of redrafting and rehearsing that focuses all our efforts on sonority and rhythm. Your text is musically structured, at least in the opening section, and I am toying with the idea of staging the whole thing in blackout, although that might be a cliché in this post-modern era, and could prove more reductive than productive. For now though, the rhythm and sound of my words for yours, and the silences they frame, have to take precedence over all other factors.

The opening section illustrates the practical difficulties facing the translator of this play into English. There is a need to find a form of language capable of marking the boundaries of a silence without disturbing it, of connoting the metaphysical dimension without denoting it, and all this in no more than half a line of text at a time. The original text (Maeterlinck 1999, p.59) reads:

Premier aveugle-né:	Il ne revient pas encore?
Deuxième aveugle-né:	Vous m'avez éveillé!
Premier aveugle-né:	Je dormais aussi.
Troisième aveugle-né:	Je dormais aussi.
Premier aveugle-né:	Il ne revient pas encore?
Deuxième aveugle-né:	Je n'entends rien venir.
Troisième aveugle-né:	Il serait temps de rentrer à l'hospice.
Premier aveugle-né:	Il faudrait savoir où nous sommes.
Deuxième aveugle-né:	Il fait froid depuis son départ.
Premier aveugle-né:	Il faudrait savoir où nous sommes!
Le plus vieil aveugle:	Y a-t-il quelqu'un qui sache où nous sommes?
La plus vieille aveugle:	Nous avons marché très longtemps; nous devons être très loin de l'hospice.

Translation draft one

1st man born blind:	Is he not back yet?
2nd man born blind:	You woke me up!
1st man born blind:	I've been asleep too.
3rd man born blind:	I've been asleep too
1st man born blind:	Is he not back yet?
2nd man born blind:	I can hear nothing coming.
3rd man born blind:	It must be time to go back to the home.
1st man born blind:	But we don't know where we are.
2nd man born blind:	It's gone cold since he left.
1st man born blind:	But we don't know where we are!
Eldest blind man:	Does anyone know where we are?
Eldest blind woman:	We were walking for ages; we must be miles from the home.

Translation draft two

1st man born blind:	Not back yet?
2nd man born blind:	You woke me up!
1st man born blind:	I too was sleeping.

3rd man born blind:	I too was sleeping.
1st man born blind:	Not back yet?
2nd man born blind:	I hear nothing coming.
3rd man born blind:	Time to go back to the home.
1st man born blind:	If only we knew where we were.
2nd man born blind:	It's gone quite cold now he's gone.
1st man born blind:	If only we knew where we were!
Eldest blind man:	Who knows where we are?
Eldest blind woman:	We walked for a very long time; we must be miles from the home.

The attempt to prioritize rhythm (often by choosing short words and phrases) in the spoken text resulted in the dramatic dialogue being kept very succinct. Metaphysical interpretations available for a theatrical reading of the French dialogue immediately receded from the English text. In the French, the use of simplistic syntax and decontextualized generic terms serves to create a poetic sense of otherworldliness by jettisoning, as it were, the linguistic clutter of everyday reality. By contrast these same strategies rendered in English seem to pull the register more towards the banal than the ethereal.

The challenge was therefore to structure the non-verbal elements of the performance text to re-engage the metaphysical resonances in a different way. A specific example of this comes later on in the play in a series of references to 'le ciel', an everyday word, which in French may carry associations of 'heaven' as well as 'sky'. The sound of wings from up above prompts questions of 'what's up there?' which of course must go unanswered in the metaphysical discourse of modernism. In the English text, to maintain the simple rhythm and generalized register we have to settle for the most innocent and mundane term, 'sky', leaving the metaphysical dimension altogether unspoken. There's a hole in the verbal text, which the staging needs to plug.

Maeterlinck attaches importance to what he calls 'second-degree' dialogue. His explanation of the term seems to exacerbate the need to maintain in performance that textual ambiguity which allows the sublime to be read through the banal, or heaven to be perceived through a reference to the sky.

Il n'y a guère que les paroles qui semblent d'abord inutiles qui comptent dans une œuvre. C'est en elles qui se trouve son âme. A côté du dialogue indispensable, il y a presque toujours un autre dialogue qui semble superflu. Examinez attentivement et vous verrez que c'est

> le seul que l'âme écoute profondément, parce que c'est en cet endroit seulement qu'on lui parle. (Maeterlinck, 1986, p.107)
> It is the words which at first seem redundant that matter in a play. It is in them that its soul lies. Side by side with the necessary dialogue there is almost always a dialogue that seems superfluous. Examine it carefully and you will see that that is the only one to which the soul will listen profoundly, because it is only here that it is being addressed. (trans. Brandt, 1998, p.119)

Ubersfeld (1999, p.15) suggests that 'the theatrical sign is flexible; it is possible to substitute a sign belonging to one code for a sign belonging to another'. The question for this project then was this: if the spoken text could not carry the multiplicity of associations in English that I perceived in the French, how else could the metaphysical be invoked in performance? We needed somehow in the process of moving from written text to performance text to privilege discourse over dramatic narrative. The symbolist aesthetic is by definition anti-mimetic, and in staging a play of this sort we needed to devise strategies that would avoid iconic representation and demand readings beyond the literal world of the story.

I will focus here on three particular strategies explored in this production process: firstly, the use of silence; secondly, the use of non-mimetic live sound, and thirdly the 'knowingness' of the actors as a way of deconstructing the dramatic representation.

In the stage directions the first sound to be heard is that of three of the women praying and lamenting constantly in muffled voices (Maeterlinck, 1999, p.58).[3] Even when they stop praying, further into the play, causing a sudden sense of rupture in the performance, there can be no 'real' silence in the performance. The presence of people in a room, including audience and actors, all agreeing not to speak, creates the most powerful experience of the liveness of performance, by the fact that it foregrounds the durational and ephemeral aspect of the performance moment. Nonspeaking is not the same as silence, however, which is the absence of sound: 'All I can hear is our breathing' says one of the characters later on and the play is full of articulations of the presence of nothingness that seem worthy of Beckett, such as 'I can hear nothing coming', in that first section. The presence of nothing was what we were seeking, the liminal tension of the twilight, not the blankness of the total dark.

Total silence in the theatre is in practical terms even more difficult to achieve than total blackout. In acknowledging the material realities of the theatre, and creating *near* darkness and *near* silence, we sought not only to invite the audience to look and to listen with greater intensity, but in so doing to recognize the intimacy of their sensory engagement with the performers in the here and now of the theatre.

In a semiological sense, 'silencing' the performance is equally impossible – the indexical signs that a performance is in progress, not least the presence of actors and audience in a designated space, cannot be switched off. However the intentional absence of speaking can *represent* metonymically total or metaphysical silence even if it cannot present it to the senses of the audience; like taste and smell and touch, silence can be virtually evoked.

In our production we attempted to exploit the particular sense of liveness that comes from tacit performance and to frame the dramatic representation with a theatrical play on silence. The knowingness with which the actors signal their performance to the audience engages a productive tension with the naivety of the blind characters they embody, in a kind of Brechtian *gestus* which makes the audience complicit with the performers in the construction of the drama and invites them to listen and look intently in active contemplation.

The actors were visible and preset around the edges of the studio while the audience entered the central space, in semi-darkness. In blackout, they performed a percussive sound sequence using found objects – material and stone, as well as various kinds of wooden blocks. As the sounds died away to leave a heavy theatrical silence that raised expectation of a text about to be spoken, the actors instead performed a semi-choreographed sequence of movements, modelled on time and motion studies, but most importantly designed to allow the corporeal presence of the actors to be experienced intimately through the rustling of their clothes as they moved almost silently in the almost dark. Relief finally comes as the first line of the dramatic text marks the first release from taciturnity.

Maeterlinck famously wanted to eliminate or at least downplay the physicality of the actor in performance. The very sound of a voice, with its unique timbre and accent, mitigates against abstraction in the spoken text: Czech structuralist scholar Jiří Veltruský (1976, cited in Pavis, 1982, p.29) describes the phenomenon as 'a dialectical tension between dramatic text and actor, a tension based above all on the fact that the acoustic elements of the linguistic signs are an integral part of the vocal resources utilized by the actor'.

Paradoxically, in Maeterlinck's theatre, it is in silence rather than speech that the text can transcend its physical embodiment. Maeterlinck has plenty to say about silence, distinguishing between 'active' and 'passive' silence and insisting that it is only in silence, not speech, that true communication may occur.

> Si toutes les paroles se ressemblent, tous les silences diffèrent, et la plupart du temps toute une destinée dépend de la qualité de ce premier silence que deux âmes vont former. (Maeterlinck, 1986, p.20)
>
> Just as all speech sounds the same, so all silences are different, and most of the time one's entire destiny depends on the quality of the very first silence that two souls will shape.
>
> *Dear Maurice,*
>
> *I've been reading up on Silence in a special issue of Performance Research. According to Rudi Laermans (1999, p.4), '[t]hat use of the stage which intentionally abuses the institutional space for the staging of a "deep" and "meaningful" quasi-mystic silence is facile. An accessory silence can only be successful if it is not rhetorical but risky.'*
> *I wonder what you'd say to that?*

Perhaps at the outset we may have been open to accusations of staging a facile rhetorical device. However, as the narrative progressed and the didascalia of Maeterlinck's text called for various diegetic sounds, such as the flapping of birds' wings, the shriek of birds, the wind in the trees, the sound of the waves against the cliffs, and the crying of the child, we were able to engage our theatrical discourse to generate a more open system of imprecise, but deliberate, non-mimetic sounds that we hoped would allow the metaphysical register to be reinstated in the reading (hearing) of the performance text. Sounds within the diegetic world, usually signalling the inhospitable natural order, would draw the listening attention of the blind characters in a series of stage silences that punctuated the rhythm of the spoken text and focused the attention of the audience to the silence in the auditorium. Susan Sontag (1982, p.191) articulates the metaphysical draw of silence in terms of a compulsion to stare:

> Traditional art invites a look. Art that is silent engenders a stare. Silent art allows – at least in principle – no release from attention, because there has never, in principle, been any soliciting of it. A stare is perhaps as far from history, as close to eternity, as contemporary art can get.

The use of indicative sound effects also afforded the spoken dialogue some level of mimetic integrity to allow it to operate within the diegetic world of the fiction. The percussion sounds were used by the actors to signal birds, waves, dead leaves and wind, while the crying of the child was created in the space by jangling bells. The closest the production came to mimetic sound was the midnight chimes of the church clock, which were produced by an actor in low light slowly striking a large triangle. The strategy established the orchestration of an intermittent dynamic soundscape within a subdued visual environment, which seemed to accord with the symbolist aesthetic.

> With their preference for voice, sound, and musical orchestration, the symbolists redirected the focus from one sense (eye) to another (ear), and further from simple sensory event to a multiple sensory experience. The emphasis on imagination, inner life, hidden reality and spiritual aspiration is directly connected to this change. (Deak, 1993, pp.176–7).

The displacement of aspects of the verbal text into theatrical and specifically paralinguistic aspects of the *mise en scène* seemed to have had a kind of ripple effect, working at the interplay of the dramatic and the theatrical, and foregrounding the experience of the theatrical event over the representation of the dramatic narrative. The intense theatricality of the event seemed to attract the ear and eye, in contemplation primarily of the aesthetic experience of the event itself and only secondarily as the representation of a significant narrative.

This foregrounding of liveness stands as an ironic inversion of Maeterlinck's famous description of a literary and logocentric theatre in which he strives to reduce as far as possible the physical presence of theatrical means, and of the human actor in particular, in favour of pure and transcendent words. However, Patrick McGuinness (2006, p.158) has demonstrated how in claiming to refuse the materiality of theatre Maeterlinck actually revalorizes it:

> In Maeterlinck's theatre, agency and expression are taken from language and its human user and transferred to the world of things: objects, sounds and off-stage space are prominent driving forces in his plays. He sought to replicate symbolist values – hiddenness, ambiguity, uncertainty – at the level of staging, set, props, and lighting, by fully mobilizing the theatre's physical resources. In so doing, he showed himself ready not just to use but to *exploit* the very 'material side' of theatre that his fellow Symbolists disdained.

The space was configured to exploit further the audience's consciousness of theatrical performance. The actors were placed initially around the outside of a single row of audience seating, arranged in an oval configuration with a central stage area visible but unlit and unused by the actors until the final moments of the play, when the woman tentatively approaches the void, led by the crying of the child. Scented lilies, with obvious funereal symbolism, were scattered across the floor. The largely unoccupied central acting area could be considered a spatial echo of Breton's literary notion of the 'full margin' which is cited by Sontag (1982, p.188) as an attempt in art to express silence through noise: 'The artist is enjoined to devote himself to filling up the periphery of the art space, leaving the central area of usage blank.'

A few wispy strips of coloured organza fabric were suspended from the grid and draped, like cobwebs or vines, into the central area. These were lit directly from above, to produce faint and slowly changing glimmers of radiance – again perhaps a suggestion of celestial light, or maybe just a splash of non-mimetic colour to please the senses and focus the mind.

Ann Ubersfeld (1999, p.10) sets out a clear model of the relationship between a written text and a performance text. She maintains that in the process of preparing a text for performance there are necessarily 'gaps' in the written text that have to be filled, and this is the job principally of the director. Patrice Pavis (1982, p.140) calls them 'holes' and essentially the lacunae consist of all the information that we need to realize the play in performance, and that is not given in the (inevitably inadequate) written text.

Dear Maurice

Just how old is an 'old woman' and what exactly do you mean by a 'madwoman'? What do the birds sound like, and how far apart are they all sitting . . . and . . . and . . . ?

PS They say you wanted to draw attention away from the physical in performance. So how on earth do we deal with the appearance of the dog in your play? And a baby?! You don't make this easy.

Ubersfeld (1999, p.10) tells us that the information we need to fill these gaps is added in some form to create another text that is not itself the performance, where the gaps will ultimately be filled, but is preparatory to it. Her model looks like this:

$$T + T' \to P$$

That T', in Ubersfeld's formulation, consists of oral and/or written material, such as director's notes or the prompt copy, amounting to a kind of written or unwritten linguistic annotation that approximates at least in function to what Pavis calls the director's metatext.

Looking at this model, I wonder quite where my translation fits in. In a traditional linear process of development from text to translation to treatment to performance, it would replace Maeterlinck's text (T1) as T2, like this:

$$T1 \rightarrow T2 + T2' \rightarrow P2$$

This is not to mention the way that any performance might itself cause revisions to the written text, creating something akin to what De Marinis (1993, p.23) calls the 'residue-text'; a text containing a transcription of performance, an a posteriori metatext. Although De Marinis states this as exceptional, he does not examine theatre translation as a specific process, which I would suggest may include more frequent occurrences of the residue-text, or at least elements of such a text, wherever a translated script, (as often now also occurs in new writing), is developed through and revised after rehearsal in the context of a specific staging.

What I have been trying to do here is to weave together the director's treatment of the source text (T') with the translation of it into English (T2) so that the two processes can both be seen as complementary aspects of the treatment preparatory to performance. So in my model, T2 and its treatment as T2' are held in relation to one another, and ideally the translated text only emerges as part of a specific treatment of the original. Perhaps it might look more like this:

$$T1 \rightarrow T1' + T2 + T2' \rightarrow P$$

The linear diagram is altogether misleading, however, because as Pavis (1982, p.6) tells us theatre semiology is *syncretic* (that is to say it puts into action many languages of expression). Therefore *mise en scène* is not just a 'straight translation' but is a simultaneous engagement of multiple 'languages' all at the same time, more of an opening 'out of – ' than a translation 'into – ', and particularly usefully for the present project to engage non-verbal performance codes into a translation and production strategy, he offers (1982, p.137) the following definition of *mise en scène*: 'activity founded on meaning-making by inter-relating heterogeneous elements'.

Pavis (1982, p.18) points out that the notion of a direct equivalence between a written text and a performance 'of' it, which lies behind the 'Page-to-Stage' model of direct transference, is misleading and reductionist, assuming as it must that the written text is the referent of the performance text.

Even among semiologists the idea still persists that the mise en scène of a text is only a transcodification of one system into another which is a semiological absurdity!

The very fact that theatrical communication is polysemic, and polyphonic, is what makes it difficult to analyse in semiotic terms, but is also what offers up a vast array of creative possibilities to the director and translator alike.

Ubersfeld (1999, p.14) makes the remarkable assertion that the multitude of codes operating in theatre 'allows even those who do not possess all of the codes to hear and understand theatre: one can understand a play without understanding its national or local allusions, or without grasping a particular complex or outdated cultural code' and cites in evidence a general ignorance of mythology, for example, amongst Racine's original audiences, declaring that it 'worked because all of the other codes in play allowed for sufficient comprehension of the signs'. Although the sense that 'it worked' in rehearsal is often intuitive for actors and director, and it is always difficult to specify the precise means by which any performance succeeds in generating a powerful connection with an audience, the vagueness of the phrase is problematic in the context of her rigorous semiotic analysis. Ubersfeld seems to assume here either duplication of signs in different codes, perhaps even of referent, what we might call theatrical overcoding, or alternatively that some signs may be dispensable and to me her statement raises complex questions about reception and the competency of the assumed spectator which are beyond the scope of this brief study.

Discussing the relationship between written text and performance text Pavis (1982, p.135) observes the tendency of text-based analyses to reduce the *mise en scène* to this notion of simple transcodification, which he contrasts with the complex functions of 'stage interpretation' and 'semiological analysis in practice':

> [M]ost of the time the analyses of the *mise en scène* are preoccupied with whether the stage 'correctly illustrates' and extends the text, whether the *mise en scène* fulfils its mission of *'theatrically'* ('expressively,' 'attractively') saying what the script has already said *textually.* [. . .] I'd like to define *the work of stage interpretation* as that of *the writing of a performance text.* (original emphases)

The spoken text in performance is only one element of the *mise en scène*, albeit the one likely to be most closely related to the written text.

Historically privileged by the notion of dramatic literature, the verbal text, comprising dialogue and stage directions, is normally considered the domain of the translator. In the present model, the translator's work extends beyond the verbal to the scenographic, visual, spatial and auditory. The need for interpretation, to govern choices in this process of expansion inevitably involves a hermeneutic as well as semiological function, and suggests a slightly revised relationship between translation and direction from the straightforward parallel that we started out with. Instead of thinking of a translation as something that generates a substitute text which in turn generates a *mise en scène*, we might think of a translation as something generated from a source text as part of a specific *mise en scène*. This model assumes the director is competent to read the original text, or that the translator and director are able to negotiate a shared reading of the original as a *mise en scène* evolves in the rehearsal process, and the treatment (T') being essential to both translation and staging processes.

Problematics of practice

The second aspect of this project explores the problematics of a process which attempts to combine the twin roles and functions of translator-director in preparing a performance text.

In terms of process, the writing of a performance text must precede performance itself, but the writing of a performance text by a director in rehearsal proceeds with a very different rhythm than the traditional process of authoring a translation. Directing is a collaborative exercise that proceeds largely in situ with a group of actors, sometimes a playwright, designer, stage manager present and contributing, by a process of experimentation, through trial and error. Translation on the other hand is usually conducted relatively slowly, in quiet isolation, away from the rehearsal room, to generate a 'product', a Text (T), which can subsequently be used to generate a *mise en scène* (T') for performance (P).

Dear Maurice,

First day of rehearsals. We started quite conventionally, with a read-through. I had drafted a translation in advance, in a fairly conventional way I suppose, trying to strike a balance between accuracy, rhythm, narrative credibility, and feeling some obligation to your literary elegance. I had left this pretty rough in most places, knowing that its purpose was to give the actors a sense of the structure and to give us something to work from, which inevitably meant it would all have to be redrafted. Although this clumsy translation was never more than a working document, it was the means by which the actors encountered your work, and in that sense I have done you a disservice and have a daunting task ahead. I think the actors felt that too.

The problem of the rhythm of the process only grew worse as time went on. I rehearsed during the day and translated into the night, feeling the pressure of the momentum of the actors' process upon the whole project. They wanted the definitive script and were understandably unnerved by a series of retrospective changes. We were all beginning to drown in a confusing sea of drafts that eventually had to be colour-coded ('We're on the Red one today, no, that's the blue one!')

The reality of a contemporary university practical timetable added a familiar sense of urgency. We had one full week in the rehearsal room to prepare the production, and so I chose to stage an abridged version (with no dog!). A rehearsed reading would have been an attractive option, but the decision to work in near darkness made reading impractical and we had to take it further, so the pressure to achieve the final script, fixed in black and white, was considerable.

Of course I had prepared a partial working draft script, thought about the range of strategies available, and tried to pre-empt problems before starting the practical work with the actors. I'd researched the play, and read Maya Slater's 1997 translation. There's only so much of a collaborative and embodied experience that a sole translator can imagine, however, and my main aim in undertaking this project was to explore the location of the translator's work in the practical and collaborative theatre-making process. David Johnston (2004, p.34) has described the invaluable experience of seeing a draft translation worked in rehearsal:

> It may be only in the rehearsal room that the translator becomes fully alive to the potentialities of performance – for better and for worse – that are encoded in his or her own playscript [. . .] Strictly speaking, they are performable, but not without striking a dissonant note within the overall orchestration of the performance.

I tried annotating and editing in the rehearsal room, but as a director I take very few notes by habit, and I soon discovered that my own physical behaviour in that role is quite distinct from, and incompatible with the sedentary activity of the writer. As a director I am always on my feet, moving about. I like my hands free and tend to locate myself physically in the space with the actors, especially in the early stages of rehearsal. As a translator, I needed to be at the desk, and didn't want to risk flattening the energy of the rehearsal by bringing yet more sheets of paper, to be studied, into the space. I struggled constantly to move between the laptop in the corner and the open space of the studio. I did not have a DSM, and I am not sure that rehearsal notes would have helped much anyway, as I would still have to redraft after the rehearsal.

The French text is formally structured along musical lines, and an apparently minor change in the translation of one line can have a series of unforeseen consequences for other moments in the play, where patterning or echoing occurs, requiring a whole series of checks and revisions, and often reversions. There is a parallel here with blocking, for example, but somehow it is easier to resolve physical and staging issues in the space with the actors, whereas textual matters require close and individual attention to the script and in rehearsal this leaves actors standing by and waiting. Moreover, in allowing a high degree of transference between signifying systems, any change in the verbal text could potentially result in another 'hole' to be filled by the *mise en scène*, and with neither the translation nor the direction clearly leading the decision-making process, there was an ever-present sense of danger of the whole thing unravelling. The constant here, to which we referred directly and frequently, was the source text, but this offered nothing that could be used by actors without intervention by the director/translator, so offered no immediate solutions.

As translator, working alone, I need to listen to the sound of the text in my own head. To me as director, the sound only becomes interesting on hearing the actors speaking it. The different habitual working modes of director and translator again generated a tension in the attempt to amalgamate the processes. Translating, I had been hearing the text in my English accent, which itself necessitated some redrafting in and after each rehearsal, to accommodate the rhythm and speech patterns of my Northern Irish actors. While easing the translation into the shape of the various accents and timbres, I was also aware of the need to resist the temptation of 'speakability' whereby difficult, formal, or non-naturalistic lines of text are softened to the pulp of contemporary everyday speech. There was an occasional tendency amongst the actors to see redrafting as an answer to the challenge of performing awkward lines, and an inclination to mistrust both the simplicity and the deliberate ambiguities of the translated text. I would be asked to make things clearer; the actors would want to embellish, to play; they would struggle with the formality of identical repetitions. *The Blind* is a difficult text to perform in the contemporary theatre. I suppose sometimes everyone wondered why as director I persisted with a highly resistant section of text instead of just asking myself for a re-draft that would be easier to play.

As rehearsals progressed I became aware that my focus was shifting, and my translation became freer, as I felt more beholden to the specifics of my actors and the particular production that was gradually forming than to the original text, that was inevitably receding. I started to recognize in my translating process Pavis' description (1982, p.205) of the director:

> The director [. . .] is someone who plays hide-and-seek with the text he claims to 'serve' [. . .] He is the man of the sign *par excellence*, because he has complete power to draw from the text, through the intervention of the stage, potentialities, possibilities, or even (sometimes) impossibilities. He plays the 'modest' role of servant to the text, yet never loses on [sic] opportunity of recalling the providential nature of his appearance.

My translation, as one aspect of my *mise en scène,* filled some holes, at the same time as it opened up others.

Dear Maurice,

Today's performance was well-received, and certainly gave the audience a rare space for contemplation amidst the hubbub of everyday life. But now I want to do it again and take it all a stage further. The words still dominate the silence, and not the other way round. I don't think I was brave enough in denying the words their privileged status, even when they had lost some of their significance, in my translation. Maybe that's a difficulty of working as translator and director - sometimes I got too comfortable with the text, or too focused on it, to really work on the unexpected in rehearsal. I also noticed how much the actors clung to the text for security, investing it with a sense of dramatic importance it rarely deserved. It's counter-intuitive to depersonalize a performance.

I want to try half-day rehearsals as well - to factor in redrafting time in the midst of rehearsals. It's not perfect, but it might work.

Don't worry, I probably couldn't publish this translation, as it's too full of holes that weren't there in your text and it needs our particular performance to plug them. But I've made a first draft of that performance, and I'm beginning to see what remains to be done.

Notes

1. In French 'l'hospice'. The word 'hospice' has in current English a rather more specific association with care for the terminally ill. While this may not be inappropriate, the rhythm, vagueness and ambiguity of 'the home' makes it preferable here in supporting the particular texture of Maeterlinck's dialogue.
2. All translations are my own except where otherwise acknowledged.
3. Everything is turned inwards, towards 'the private inner life of the soul'. One of the women is mute ['muette'], their voices are stifled ['sourdes'] and of course they are all blind ['aveugles'] – in their isolation the innocents of this play hear, see and speak no evil.

References

Banes, S. and Lepecki, A. (eds.) (2007) *The Senses in Performance* (London and New York: Routledge)
Boswell, L. (1996) 'Interview. The Director as Translator', in D. Johnston (ed.) *Stages of Translation* (Bath: Absolute Classics), pp.145–52
Brandt, G. W. (ed.) (1998) *Modern Theories of Drama: a Selection of Writings on Drama and Theatre, 1850–1990* (Oxford: Clarendon Press)
Deak, F. (1993) *Symbolist Theater: the Formation of an Avant-Garde* (Baltimore: Johns Hopkins University Press)
De Marinis, M. (1993) *The Semiotics of Performance*. A. O'Healy (trans.) (Bloomington: Indiana University Press)
Déprats, J. M (ed.) (1996) *Antoine Vitez: le devoir de traduire* (Montpellier: Editions Climats et Maison Antoine Vitez)
Johnston, D. (2004) 'Securing the Performability of the Play in Translation', in S. Coelsch-Foisner and H. Klein (eds.) *Drama Translation and Theatre Practice*, Salzburg Studies in English Literature and Culture, 1 (Frankfurt am Main: Peter Lang), pp.25–38
McGuinness, P. (2000) *Maurice Maeterlinck and the Making of Modern Theatre* (Oxford: Oxford University Press)
McGuinness, P. (2006) 'Mallarmé, Maeterlinck and the Symbolist *Via Negativa* of Theatre', in A. Ackerman and M. Puchner (eds.) *Against Theatre: Creative Destructions on the Modernist Stage*, (Basingstoke and New York: Palgrave Macmillan), pp.149–70
Maeterlinck, M. (1986) *Le Trésor des humbles* (Bruxelles: Editions Labor). Extracts from this text translated by the editor in Brandt (1998)

Maeterlinck, M. (1997) *The Blind (Les Aveugles)* in M. Slater (trans. and ed.) *Three Pre-Surrealist Plays* (Oxford: Oxford University Press, World's Classics Series), pp.1–48

Maeterlinck, M. (1999) *Les Aveugles,* in Maeterlinck, *Pelléas et Mélisande, Les Aveugles, L'Intruse, Intérieur,* L. Hodson (ed.) (London: Bristol Classical Press, French Texts Series), pp.58–78

Pavis, P. (1982) *Languages of the Stage: Essays in the Semiology of the Theatre,* 2nd printing (New York: Performing Arts Journal Publications)

Sontag, S. (1982) 'The aesthetics of silence', in *A Susan Sontag Reader* (Harmondsworth: Penguin Books), pp.181–204

Stafford-Clark, M. (1989) *Letters to George* (London: Nick Hern Books)

Ubersfeld, A. (1999) *Reading Theatre,* F. Collins (trans.) (Toronto: University of Toronto Press) [Translation of 1996, *Lire le théâtre*]

Upton, C.-A. (2000) *Moving Target: Theatre Translation and Cultural Relocation* (Manchester: St Jerome)

Zuber-Skerritt, O. (1984) (ed.) *Page to Stage: Theatre as Translation* (New York and Amsterdam: Rodopi)

3
Musical Realizations: a Performance-based Translation of Rhythm in Koltès' *Dans la solitude des champs de coton*

Roger Baines and Fred Dalmasso

Introduction

This essay provides an account and analysis of an experimental method focusing on the performance of rhythm and sonorities to produce a new translation into English of *Dans la solitude des champs de coton* (1985) by Bernard-Marie Koltès (1948–89). Koltès' works have continued to grow in popularity in Europe since his death in 1989 and he is now considered one of the most original and influential French playwrights of recent decades. *Dans la solitude des champs de coton* (hereafter *Solitude*) is an encounter between two characters, a Dealer and a Client, in an ill-defined nocturnal space. On his way from A to B, from one lit window at the top of a building to another, the Client enters the territory of the Dealer. They circle around each other for the duration of the play, but the Client will not reveal what he desires and the Dealer will not reveal what he has to offer to satisfy the Client's desires; the play unfolds until it ends in a moment of inconclusive potential violence.

The reception of Koltès' work on the English-language stage has been identified as problematic. According to Maria Delgado and David Fancy (2001, pp.149–50), this may in part be due to the emphasis in English-language theatre on psychologically driven characters and action-centred narrative; but we believe that it may also be because existing translations do not completely engage with the fundamentally important rhythmic qualities of Koltès' writing; certainly we feel that this aspect of his work in translation deserves more attention.

Koltès (1999, p.27) writes in his collection of autobiographical interviews *Une part de ma vie*:

> The French language, like French culture in general, only interests me when it is distorted. A French culture, revised and updated, colonized by a foreign culture, would have a new dimension and its expressive riches would be increased, like an ancient statue with no head or arms which draws its beauty from this very absence.[1]

Solitude is a particular text because of the way in which the playwright uses and distorts complex syntactical structures and seemingly fixed rhythmic patterns specific to the French language, but also because of the way in which Koltès deploys sonorities and rhythm to make such a multi-layered text performable.

The following analysis is the product of a reading of the text which is, crucially, influenced by performance. Our approach to stage translation emphasizes the adaptation of the rhythm of the source text to the target language thus allowing the original rhythm to be retained. Rhythm, however, is not dealt with as inert and inflexible but as malleable and adaptable so that the original rhythm is allowed to survive the translation process. The idea is to create a musical variation of the original text. For example, we did not try to either recreate the deviations from formal structures or to repair cohesion or meaning. We sought a text in translation that had a similar trajectory to the original, a momentum provided by the complex structures and the minimal use of full stops. As stated above, importantly, our method of achieving this was the development of a translation in conjunction with a performer who is expert in performing and designing complex texts for speech, and can locate appropriate rhythms and sonorities in the English text via speech, rhythmic vocalization and drumming. We thus set ourselves the problematic of whether we could successfully establish rhythm in sound, and tempo through alliteration and assonance, and whether the choices of syntactic structures and lexis go hand in hand with the physical and vocal rhythms of the performer. Consequently, our method is one which transfers the text from one performance to another and experiments with a reading of a play which is informed by translation practice.

The rhythm of Koltès' writing in *Dans la solitude des champs de coton*

A dictionary definition of rhythm gives us 'an effect of ordered movement in a work of art, literature, drama, etc. attained through patterns in the timing, spacing, repetition, accenting, etc. of the elements' (Websters, 2009). In opposition to this common conception of rhythm as a regularity of similar intervals or recurrences, based on repetition, periodicity and measure, French linguist and translator Henri Meschonnic defines rhythm as a disposition or configuration without any fixedness, one which results from an arrangement which is always subject to change (Meschonnic, 1982). So, how is rhythm achieved in Koltès' texts? How does he enable his complicated texts to flow in performance, giving the actor 'rails' on which to run? As mentioned above, what is distinctive in Koltès' writing in this play is its apparent linguistic formality. This is achieved in part through the use of a text which, to an extent, subverts classic French alexandrine structures. It is also achieved through the use of complex syntax with minimal punctuation, described by a critic of Patrice Chéreau's 1995 surtitled Edinburgh production as 'mind-numbing sentences that zigzagged through ranks of subordinate clauses' (Delgado and Fancy, 2001, p.155). This description of one translation of the text gives an idea of some of the challenges that it poses to translators. The following, more detailed, analysis of the particular features of Koltès' text which are especially relevant to our translation process will enable us to present an assessment of the issues on which we elected to focus our experimental translation strategy. This strategy will then be discussed and illustrated.

The apparent formality of the text is periodically punctured when stanzas based on an elusive twelve syllable alexandrine are interrupted or distorted, when highly complex syntax is stretched to extremes. The syntax accommodates the stanzaic pattern but also erects obstacles in the flow of the text. As the text is spoken, an irregular dodecasyllabic metric system evoking the alexandrine appears and disappears throughout the text but imprints a durable rhythm. The Dealer sets the rules for the ensuing verbal jousting, as if he is laying down a rhythm which needs navigating back and forth from, and the Client follows the same cadence, the same rules, as though in a competition. If communication is present, it is enacted through rhythm, the two protagonists jousting with language to attempt to unseat the opponent, and this is indeed one of the images the Dealer uses. Like the Dealer who reins in his tongue so as not to unleash his stallion, the characters seem to be chasing rhythm through the text. The elusive alexandrine can be identified as the basic unit of organization and the frame within which acoustic designs are composed.[2] The music of a sequence of lines is to be found in the rhythmic movements back and forth from the haunting alexandrine cadence. The alexandrine verse is the most deeply ingrained melodic structure in the French language, it is the most suited to the melodic movements of the language.³ Given the length and complexity of the sentences in *Solitude*, the performer might indeed fall back upon the natural rhythm of the alexandrines for his delivery and somehow force them upon the text. The way in which Koltès' text evokes the alexandrine has parallels with Schoenberg's conception of atonal music as a free, twelve-tone chromatic field where any configuration of pitches could act as a 'norm', a flexible way of providing a foundation with any number of variations on that foundation.

Like the alexandrines, the syntax in the play is overarching but it often carries so many interpolated clauses governed by the same main verb (which, in some cases, is so distant that the interpolation is isolated) that it appears to lose grammatical coherence. In fact the text is so well crafted that, although the syntax is stretched to the limit, there is nonetheless grammatical coherence. As with the alexandrine structure, the syntactic cohesion is omnipresent but not always foregrounded. The following is an example of a highly formal structure which sets up a complex syntactical shape that is stretched to the limit:

A
Le Dealer
Alors ne me refusez pas de me dire l'objet, je vous en prie, de votre fièvre, de votre regard sur moi, la raison, de me la dire; et s'il s'agit de ne point blesser votre dignité, eh bien, dites-là comme on la dit à un arbre, ou face au mur d'une prison, ou dans la solitude d'un champ de coton dans lequel on se promène, nu, la nuit; de me la dire sans même me regarder. (2004, p.31)

So do not refuse to share the object, pray, of your fever, of your eyes on me, the reason, share it; and if that compromises your dignity, well then, say it like you would say it to a tree, or to a prison wall, or in the solitude of a cotton field where you would wander in the nude, at night; share without even a glance. (Baines and Dalmasso, unpaged)

Here, there is an elliptic construction ['de me la dire'] which is used chiastically as the referent for the final phrase of the sentence. In the end, the Dealer would be satisfied with any response from the Client. The request in the sentence becomes increasingly abstract: first the Dealer asks for the object of the Client's desire, then more abstractly the reason for this desire, and then he reiterates his question resorting to the pronoun 'la' in place of 'la raison', but pronoun and noun are so distant in the sentence that the object of the question is almost forgotten.

All that remains is the prompt to speak, 'de me la dire'. The text seems here to exhaust both syntax and meaning.

Next, a different example of complex syntax which, although generally more typical of French than English, is nonetheless here pushed to extremes:

B
Le Client
Et si je suis ici, en parcours, en attente, en suspension, en déplacement, hors-jeu, hors vie, provisoire, pratiquement absent, pour ainsi dire pas là – car dit-on d'un homme qui traverse l'Atlantique en avion qu'il est à tel moment au Groenland, et l'est-il vraiment? Ou au cœur tumultueux de l'océan? – et si j'ai fait un écart, bien que ma ligne droite, du point d'où je viens au point où je vais n'ait pas de raison, aucune, d'être tordue tout à coup, c'est que vous me barrez le chemin, plein d'intentions illicites et de présomptions à mon égard d'intentions illicites. (2004, p.19)

And if I am here, on the way, on the move, paused, postponed, out of sync, out of joint, provisional, practically absent, say, not here – would you locate a man flying across the Atlantic at such and such a moment in Greenland, is he there? Or right in the tumult of the ocean? Say I did step aside, although my straight line, from point of departure to arrival has no reason, none, for a sudden curve, well, you are blocking my path, full of illicit intentions and misplaced assumptions of my illicit intentions. (Baines and Dalmasso, unpaged)

The bare bones of the sentence are 'si je suis ici [. . .] et si j'ai fait un écart [. . .] c'est que vous me barrez le chemin [. . .]' (if I am here [. . .] say I did step aside [. . .] well you are blocking my path), yet there are as many as seventeen diversions from this simple premise in the one sentence. Foregrounded in the structure of the complex sentence in the example above is 'the isolation of the man in the plane lost in the ocean', a metaphor for the construction of a text which isolates clauses – in this case the very clause which is the metaphor – in the tumult of the syntax, and this is a distinctive feature of a text which has performative features, a text which often 'does what it says'. The performativity of the language used in the text works on a number of levels, it provides metaphors for the situation of the characters and isolated sounds which perform actions and interrupt the flow of the text.

The nature of the syntax, and indeed the architecture of the play as a whole, enable it to be read as a work which moves towards distillation, then towards a void, and it is this interpretation which led us to elaborate a particular translation method with its focus on rhythmic qualities, on sound and melody. Such is the strength of the rhythmic structure that it imposes itself on the audience and creates an expectation. In many ways, this strong structure enables Koltès to write more freely, akin to a traditional jazz piece, where more risks are taken melodically in the music at certain stages within a structure. This is an important metaphor which we will return to later. The performative metaphor in the example above also functions as a metaphor for the situation of the characters in the play who are lost on a trajectory between two points of light. The Dealer's speech has what could be called 'motionless movement' because he always returns to the same point, the same question: 'what is your desire?' The Client, in contrast, wants to pass through:

> et la ligne droite, censée me mener d'un point lumineux à un autre point lumineux, à cause de vous devient crochue et labyrinthe obscur dans l'obscur territoire où je me suis perdu. (2004, p.20)

> and the straight line, there to lead me from one point of light to another point of light, because of you, becomes crooked and a dark labyrinth in the dark territory where I have strayed. (Baines and Dalmasso, unpaged)

He is more straightforward but he stumbles on the Dealer and from then on their speeches are mutually imbricated. In their own speech they single out fragments from each other's speech and use them in their own speech. In turn this repetition provides a kind of coherence and rhythm within the whole text because they always return 'à cette heure et en ce lieu' (2004, p.9), at the same hour to the same place, to each other's phrases. The rhythm of the text is both created and distorted on a macro level by the to-and-fro motion of the alexandrines and the use of highly sophisticated syntax. On a micro level, rhythm is both created and distorted by the use of sound, which furnishes another aspect to the performative character of the text. Isolated sound clusters appear, what one might call sound buttresses, which contribute to and support the overall rhythmic structure but also go against the flow, operating as counterpoints. The heavy use of the relative pronouns 'que', 'qui', 'quiconque' in the opening lines are examples of isolated sounds which contribute to the foundation of the rhythm, the repetition of the sound

[k] (an obstruent voiceless velar stop, or a sound formed by *obstructing* outward airflow) functions like a thumping drum beat. This sound also evokes that of an animal restrained by a lead and foresees the action, performed by the text, of an animal savagely baring its teeth. The final sound of the Dealer's speech 'les dents' is isolated and goes against the flow of the rhythm of the sentence but it also provides an anti-cadence, a sound which stands alone at the end of the sentence and, as such, performs.

C
Si vous marchez dehors, à cette heure et en ce lieu, c'est que vous désirez quelque chose que vous n'avez pas et cette chose, moi je peux vous la fournir; car si je suis à cette place depuis plus longtemps que vous et pour plus longtemps que vous, et que même cette heure qui est celle des rapports sauvages entre les hommes et les animaux ne m'en chasse pas, c'est que j'ai ce qu'il faut pour satisfaire le désir qui passe devant moi, et c'est comme un poids dont il faut que je me débarrasse sur quiconque, homme ou animal, qui passe devant moi.
    C'est pourquoi je m'approche de vous, malgré l'heure qui est celle où d'ordinaire l'homme et l'animal se jettent sauvagement l'un sur l'autre, je m'approche, moi, de vous, les mains ouvertes et les paumes tournées vers vous, avec l'humilité de celui qui propose face à celui qui achète, avec l'humilité de celui qui possède face à celui qui désire; et je vois votre désir comme on voit une lumière qui s'allume, à une fenêtre tout en haut d'un immeuble, dans le crépuscule; je m'approche de vous comme le crépuscule approche cette première lumière, doucement, respectueusement, presque affectueusement, laissant tout en bas dans la rue l'animal et l'homme tirer sur leurs laisses et se montrer sauvagement les dents. (2004, pp.9–10)

If you are out walking, in this neighbourhood, at this hour, you must desire something you're missing, that something I can supply; the reason I've been here for longer than you and will stay longer than you, and the reason, this very hour, the coarse intercourse of men and animals does not cause me to run, is that I have the wherewithal to satisfy the desire which passes by me, a weight of which I need to relieve myself upon anyone, man or animal, who passes by me.
    This is why I bring myself closer to you, despite the hour which normally sees man and animal in a savage rush at each other's throats, I bring myself closer to you, with my hands open, palms towards you, with the humility of he who has something to offer

up against he who has a need to buy, with the humility of he who owns, up against he who desires; and I can see your desire as you see a light go on, at the highest window, right at the top, at sundown; I bring myself closer to you like the sundown comes close to that first light, soft, respectful, almost with affection, and leave the street down below to animal and man to pull on their leads and savagely bare their teeth. (Baines and Dalmasso, unpaged)

Other examples of the original text being performative abound and we will return to this aspect shortly, but first the text has a final characteristic which enables Koltès to facilitate the delivery of the distorted alexandrines and complex syntax and thus to provide smooth 'rails' for a performer to run on, and these are what could be called sound patterns. Such sound patterns facilitate the performative function of certain lines and also provide the performer with indications of where to breathe. In the characters' cat-and-mouse play, alliterations are either free to run or stopped short. The following is a perfect illustration of the correlation between action and language in the play:

D
mais ne me demandez pas de deviner votre désir [. . .] (2004, p.12)

but do not demand of me to divine your desire. (Baines and Dalmasso, unpaged)

The Dealer is constantly trying to stop the Client from leaving and his only weapon is language, so the delivery in French of 'mais ne me demandez pas de deviner votre désir' is that of a man in such a hurry to get his words out that he produces the sound of someone stuttering in haste. A more complex illustration of the use of sounds comes in the following sentence from the Client:

E
Mon désir, s'il en est un, si je vous l'exprimais, brûlerait votre visage, vous ferait retirer les mains avec un cri, et vous vous enfuiriez dans l'obscurité comme un chien qui court si vite qu'on n'en aperçoit pas la queue. (2004, p.15)

My desire, were there one, were I to divulge it, would burn your face, have you retract your hands with a scream and you would flee into the dark like a dog who runs so fast his tail fades away. (Baines and Dalmasso, unpaged)

This sentence accelerates like the dog which is used as a simile at the end. All of 'il en est un', 'si je vous l'exprimais', 'brûlerait votre visage' and 'les mains' provide a series of closed, muffled sounds; they are then followed by the much more incisive 'avec' and the acute and open sound of 'cri' which opens the mouth, ready to accelerate the tempo in a series of brief sharp sounds, 'en-fuir-iez', 'obscurité', 'chi-en', 'qui court', 'si vite', which quicken the pace considerably. The sentence then slows down: 'qu'on n'en aperçoit pas' before spitting out the final two syllables, 'la queue'.

This example is a very pertinent one as far as our translation strategy is concerned. The final sound, 'la queue' is again separated out (like 'les dents') which draws attention to the disappearing tail, but it also provides us with a significant metaphor for the way in which the syntax operates throughout the play. All the sentences point towards a disappearance into the dark, just like the dog's tail. The syntax functions as an allegory of the story (two people meet and exhaust all they have to say and then leave or kill each other and what remains is silence). The play begins to distil, to boil down from the point in the text when the Client briefly shows some vulnerability, comparing his situation when the Dealer came across him at the start of the play to that of himself like a child in bed who cries out when his light goes out ('comme un enfant dans son lit dont la veilleuse s'éteint' [2004, p.34]). The Client starts to lose his breath out of fear and the speeches become progressively shorter. It is the apex, the summit of the play which accelerates its distillation from then on. The necessary syncopated respirations, which ensured successful negotiation of the complex syntax, become interspersed silences. The play, which follows the classical rules of unity of space, time and action, resorts to another classical verse drama device: stichomythia as the dialogue becomes tighter and tighter and alternates more rapidly from here on. This is the final element of the performative nature of this text, the way in which the micro and macro rhythms detailed above contribute to the text's dissolution, its acceleration towards the void, towards the one word penultimate speech of the play where the exhaustion of meaning and syntax evident in the 'de me la dire' example above reach their natural conclusion in the Dealer's 'Ri-en'(nothing) (2004, p.61) which could evoke the snarling of animals at the very point when all human discourses have been exhausted.

Elements of Richard A. Rogers' 1994 essay on rhythm as a form of discourse that is central to social organization, are useful to our analysis. He discusses the power of entrainment – the locking up, or synchronization, of different rhythmic patterns when they are placed in close proximity – in channelling and coordinating human energies. He uses the Grateful Dead percussionist Mickey Hart's accounts of drumming activities for children at a summer camp to further illustrate this entrainment (Hart, in Rogers, 1994, p.238):

> It's interesting how long it takes people to entrain. These kids locked up after about twenty minutes. They found the groove, and they all knew it. You could see it in their faces as they began playing louder and harder, the groove drawing them in and hardening. These things have life cycles – they begin, build in intensity, maintain, and then dissipate and dissolve.

The description of a rhythmic intensity that then dissipates and dissolves is very similar to processes we have shown are in operation in Koltès' text, while the focus on the use of music and of drums to create rhythm bears similarities with the creative processes involved in our own translation. Rogers (1994, p.362) goes on to establish the vast differences between Western and West African polyrhythms via an analysis of the speed and time-cycles of the latter's musical rhythms and makes the point that 'African rhythms are not only multiple but incomplete'. He also uses John Miller Chernoff's analysis of African rhythm and African sensibility in which Chernoff (1980, pp.113–14) makes the point that:

> The music is perhaps best considered as an arrangement of gaps where one may add rhythm, rather than as a dense pattern of sound. In the conflict of the rhythms, it is the space between the notes from which the dynamic tension comes.

The reducing trajectory of Koltès' text takes this spacing further as it leads in the end to a conflictual silence between the characters. However, like the syncopated rhythms of reggae music, for example, where it is the gaps between the beats which provide the rhythm, it is also the overarching rhythmic structure of stretched syntax and distorted alexandrines in Koltès' text which makes space for the unsaid to be heard. The audience is compelled to fill these gaps and silences. Absence, disappearance and silence generate meaning: 'like an ancient statue with no head or arms which draws its beauty from this very absence' (Koltès, 1999, p.27). The distillation of the text of *Solitude* towards the final call to arms: 'So which weapon?' dissolves meaning into pure violence. That passages of Koltès' texts perform actions and have their emphasis on sound, indeed on the display of meaning in sound, is a theme which runs through his work and further inspires our own creative work on sound and our attempt to replicate the distillatory trajectory of the text in our overall translation strategy. Translation is a way of managing spaces to enable exchange. By giving space to the text, the translator makes audible what is unsaid in the original, audible because translating plays is also conveying what is unsaid, what is not text. If you add to the text, through expansion for example, then there is simply no space for what is between the words to be heard. 'Ri-en' ('none'), uttered at the end of *Solitude*, could evoke the snarling of animals at the very point when all human discourses have been exhausted.

In the process and product of the translation we wanted to be sensitive, as far as possible in English, to what we identified as the function of the complex syntax, the flow of the text and the use of particular isolated sounds or silences to provide counterpoints to and syncopate this flow, the use of sound patterns more generally and, in conjunction with this use of sound, the performative nature of lines. The very strong distillation, and indeed acceleration, provided by the text's rhythmic architecture led us to decide that our translation strategy needed, above all, to be one which retained the spaces between the text, between the metaphorical notes.

The musical metaphor can be taken further if we consider the text to be a score whose musicality we emphasize in translation. The complex syntax is carefully constructed and functions like a musical score which returns to a central key, the main clause in each sentence, while the various parentheses and relative clauses explore the different notes of the scale. The complex syntax functions like a theme in jazz which the various improvized solos are connected to, adding sound patterns similar to those illustrated above which stretch the theme to its limits but always follow the rhythm. The arrangement of the whole piece is left to the conductor/translators. In translation, we sought to create new realizations of the piece, musical interpretations of the original.

Translation

Our concern to retain as much as possible the complex syntax, the virtuosity of the piece, meant that the word order and minimal punctuation of the original were adhered to as much as possible. The source text's complex structures keep potential strands of meaning as open as possible and retaining these structures in translation enabled us to better sustain the complexity of syntax in a performable way by maintaining a system of stratified clauses as seen in example C. Consequently both texts demanded to be looked at in other terms, in terms of rhythmic qualities and in terms of sonority, other ways of evoking meaning. In translation, we sought a balance between regular metrics and an aleatory rhythm. It is only by reworking the rhythm first and refraining from making assumptions about meaning (and thus closing down possibilities of meaning) that words will be allowed to constantly re-arrange themselves for the audience. This essay does not deal with translation and music per se. However, if we consider Franzon's proposal (2008, p.376) of a spectrum of five choices in song translation which runs from 'translating the lyrics but not taking the music into account' to 'adapting the translation to the original music', our method is closest to the latter. As mentioned above, we did not try to either recreate the deviations from formal structures or to repair cohesion or meaning. We sought a text in translation that had a similar trajectory to the original, a momentum provided by the complex structures and the minimal use of full stops. Care was taken to render these complex sentences more performable by paying attention to both the flow of the text, via the strategy applied to the translation of explicative conjunctions, and the avoidance of spelling anything out in the interests of cohesion, but also via the choice of sounds, trying to establish rhythm in sound above all, retaining the tempo mainly via alliteration and assonance.

Given our overarching concern with mirroring the distillation of the text, with retaining the gaps between the metaphorical 'notes', the one translation strategy that we could not allow ourselves to employ was that of addition and explanation, what, in Antoine Berman's list of the deforming tendencies of translation (2004, p.282), would come under the headings of 'clarification' and its consequence, 'expansion' which he describes both as 'addition which adds nothing' and as 'a stretching, a slackening, which impairs the rhythmic flow of the work'.[4] For example:

F
comme, au restaurant, lorsqu'un garçon vous fait la note et énumère, à vos oreilles écoeurées, tous les plats que vous digérez déjà depuis longtemps. (2004, p.14)

as in the restaurant, when the waiter draws up the bill and enumerates, for your queasy ear, all the dishes you have long since digested. (Baines and Dalmasso, unpaged)

Or if we take example E quoted above:

Mon désir, s'il en est un, si je vous l'exprimais, brûlerait votre visage, vous ferait retirer les mains avec un cri, et vous vous enfuiriez dans l'obscurité comme un chien qui court si vite qu'on n'en aperçoit pas la queue. (2004, p.15)

My desire, were there one, were I to divulge it, would burn your face, have you retract your hands with a scream and you would flee into the dark like a dog who runs so fast his tail fades away. (Baines and Dalmasso, unpaged)

As already described above, this sentence accelerates like the dog which is used as a simile at the end, and we were careful to retain this gradually accelerating tempo in our translation. Particular work on sound and 'musical arrangement' is in evidence here as well, and we will examine this presently, but this extract is also an example of our translation retaining this kind of accelerating movement in the text which is necessary to replicate the accelerating movement of distillation towards what then becomes a void at the end of the play.

This strategy of avoiding additions as far as possible is also relevant to the way in which we approached the complex syntactical structures and is most evident in our approach to rhetorical explicative conjunctions such as 'c'est que' (it is that) or 'c'est pourquoi' (this is why) and the ensuing interpolated clauses. Anyone who translates between French and English will know that French can sustain cohesion over much longer stretches of language than English can; a translator is often required to deal with the complex syntax of French by splitting sentences and repeating the subject, or by adding in explanatory phrases or additional relative clauses that provide the necessary cohesion in English. However, here, such explanatory additions destroy rhythm and thus make the text much harder to perform. More fluid ways of constructing such complexity needed to be found, as an analysis of the opening speech of the play shows:

If you are out walking, in this neighbourhood, at this hour, you must desire something you're missing, that something I can supply; the reason I've been here for longer than you and will stay longer than you, and the reason, this very hour, the coarse intercourse of men and animals does not cause me to run, is that I have the wherewithal to satisfy the desire which passes by me, a weight of which I need to relieve myself upon anyone, man or animal, who passes by me. This is why I bring myself closer to you, despite the hour which normally sees man and animal in a savage rush at each other's throats, I bring myself closer to you, with my hands open, palms towards you, with the humility of he who has something to offer up against he who has a need to buy, with the humility of he who owns, up against he who desires; and I can see your desire as you see a light go on, at the highest window, right at the top, at sundown; I bring myself closer to you like the sundown comes close to that first light, soft, respectful, almost with affection, and leave the street down below to animal and man to pull on their leads and savagely bare their teeth. (Baines and Dalmasso, unpaged)

The initial conjunction, 'c'est que' has been elided. For the next 'c'est que', starting the clause with a noun ('the reason is') enables us to launch the ensuing clause with a simple noun ('the reason') rather than the rhythmically destructive 'because'. The conjunctions 'parce que' (because), 'c'est pourquoi' (this is why), 'c'est que' (it is that), 'non pas que' (not that), 'si' (if/while) or 'car' (as) throughout the text invite the expectation that meaning is about to be laid bare, that clarification is imminent. There is a tension in the text between the high number of these pronouns and the few occurrences in comparison of hypothetical structures such as '[. . .] si par hypothèse je vous disais [. . .]' ('[. . .] if, just a hypothesis, I were to say [. . .]') (2004, p.42). The attempt by the characters to make sense of the situation and establish cause and effect is cancelled out by the very abundance of these conjunctions. The interpolated clauses betray incoherence by excess and thus increase possible interpretations. Reproducing the abundance of explicative conjunctions in English would lead to over explication and narrow down the pathways of meaning and so was avoided. A significant additional benefit is that not resorting to too many explicative conjunctions enabled us to avoid using the harshest sounding options in English and thus avoid interrupting rhythm. Our strategy is the inverse of Koltès' as we juxtaposed clauses diluting the use of the explicative conjunctions; but we linked the juxtaposed clauses through sound and rhythm, thus creating an apparent unity of discourse which also points to the lack of correlation between clauses and thus the inability of the characters to make sense of their situation. Throughout the translation process, we were constantly rejecting 'because' as a translation of 'c'est que', 'c'est pourquoi' or 'car', or even 'which' for 'que' because of the way in which, in English, for us in this context, they also interrupted the rhythm we were endeavouring to create. A similar strategy was applied to adverbial phrases, the '–ly' endings of 'slowly', respectfully' and 'affectionately', and the 'ing' endings of present participles (see 'I'm approaching you' versus 'I bring myself closer to you' above) because both the adverbs and the present participles bring in an idea of duration, in meaning as well as literally taking longer to say in sound, and thus tended to entangle, or even strangle, the rhythm to which we were trying to give a voice. Thus, we developed a strategy of constructing phrases that enabled us to retain the complexity of the original without the accumulation of the harsh sounding and rhythmically destructive English hinges of 'because' or 'which' and the endings of adverbs and present participles. The higher propensity of English to use intransitive verbs which require prepositions posed a similar problem because using them meant that both meaning and rhythm would be interrupted and deferred. For example, the use of the verb 'to help' in the following phrase from Wainwright's translation 'that something I'm sure I may help you with' (2004, p.91) compared with our translation of 'that something I can supply'. The analysis above explains how we approached the translation of the musical theme, of the complex syntax. The translation of the elusive/hidden alexandrines, of the rhythm section, however, posed a different problem and led to a different solution.

We have used the theme of jazz as our metaphor for the rhythmic function of the echoing alexandrine stanza, but, as mentioned above, the way in which Koltès' text evokes the alexandrine also has parallels with Schoenberg's conception of atonal music. Counterpoint (simultaneous combination of two or more melodies, one being the counterpoint of the other) is a key part of this atonal system, the structure is over-arching, sometimes prominent, sometimes in recess. The distinctive twelve-timbre rhythmic structure of the alexandrine does not exist in English and so our solution was to draw on different rhythmic structures, those of Hip Hop and slam/jazz performance poetry.

Process and performer

This musical approach obliged us to continually rehearse the translation as it evolved. Our translation process involved first performing the lines in French to enable us to locate the rhythms and then producing an initial version. In English we did not have the rhythms of the French alexandrines and so, rather than a classically trained actor, we needed a physical performer to facilitate and help us to elaborate the delivery of the sound structures we had created. Performance poets produce and *perform* difficult texts which have their foundation in sound and it is the theatrical nature of performance poetry, especially slam poetry, which is crucial here, the way in which it animates text. We attended a number of the Blackdrop performance poetry events which have been running in Nottingham since 2002 and engaged one of the founder members and regular performers, British Bajan jazz poet and musician David 'Stickman' Higgins, to test out our texts, to help us to refine the rhythms we had created in the target text.[5] We felt that this was an appropriate choice, given Koltès' interest in African and Afro-Caribbean culture and music, and the fact that his writing is influenced by the journeys he took which placed him in contact with African culture.[6] In attempting to establish a link between the performer and the performance style we used, it is worth clarifying that, although slam poetry in particular has its origins in US black culture, what we used was black British slam/jazz poetry which similarly has its origins in US hip hop culture but also has strong Afro-Caribbean influences, as does our performer. We needed someone who had the ability to tap into the rhythms of slam, rap, hip-hop and reggae but also who was accomplished enough to escape these rhythms as well.

Stickman's role was to help us to refine the rhythms and sonorities we had created. Stickman's first sight of the text was its projection onto a large screen, thus freeing his movement. Dalmasso performed the text in French so that Stickman, who does not speak French, could hear the rhythm without the distraction of meaning. He could hear where the construction of the strata of the text guides the breathing in such a manner that it cannot be performed in a wide variety of ways, especially at full speed, and then Stickman read/performed our translation. The only advice that Koltès ever gave to actors of his work, as reported by Bruno Boeglin (2001, p.46) was that they should perform as if they had an urgent need to urinate. This constraint (not particularly difficult as his natural performance style involves considerable speed) enabled Stickman to find rhythm and melodies in the text. Such a delivery suits this text in particular because the sparring characters are doing all they can to keep the opponent on the back foot, and slam poetry in particular is often a competitive art, performed in the context of slam contests which further suits our choice of performer and performance style. As a musician (his motto is 'I write, I drum, I drum, I write'), Stickman searched for a rhythm both on his drum and vocally. When he was recreating Dalmasso's delivery of Koltès' lines we noticed he employed rhythmic scaffolding derived not only from his individual voice and body but also from vocal melodic and physical patterns which could be identified as those used in hip hop, reggae and slam. By performing, drumming and vocalizing the cadences of the original, we tried to identify a musical variation in English on the rhythmic themes developed by Koltès. We found it especially useful to expose the performer to the source language, to the rhythm of the original text. Stickman could assess whether the musical variation produced worked, rhythmically and musically, like assessing a cover version of a jazz tune. While it is, of course, subjective to assess someone else's rhythm and transpose it into a musical score, the advantage of identifying a rhythm like this is that it enables us to assess our translation process in a tangible way. It is both Stickman's vocal performance and physical performance which interest us; both convey a very particular rhythm. This was entirely appropriate because it was a performance feature which echoed the self-imposed frame Koltès uses in his texts with the alexandrines. Koltès' use of sound patterns helps facilitate the delivery of the complex syntax because they help to structure the performer's breathing and therefore his rhythm.

The text requires high levels of concentration by performers because they need to be alert to each other's delivery in order to enact the complexity of the rhythmic and syntactic structures. The text provides steps which must be negotiated; it necessitates a delivery which has a rising and falling pitch. The sentence in example B above with seventeen diversions from the main premise is a good case in point, as is the start of the play, example A. Koltès' text needs measure and breathing which is helped by the sound patterns he creates and the strata of the complex syntax. In order to provide similar measure and breathing in English we also need the repetitions, the structures but, in the absence of the alexandrines, we needed to emphasize the sounds much more in order to give the text more scaffolding against which we could prop the different strata.

This is where we, as translators and performers, fit into the musical metaphor: our emphasis on sound, to some extent over meaning as we will discuss below, gave us the role of musical interpreters, instrumentalists, working out a variation on an original. The role of Stickman as the performer in this process was crucial, as was the role of Dalmasso as a performer of the French text, since within the translation process there was effectively a translation from one performance to another.

Speed is the essence of the delivery of Koltès' text in this play; it guarantees a safe navigation between the text's different levels. The image of silent film star Harold Lloyd walking on a building site and inadvertently stepping on steel girders being lifted by cranes but managing to follow an uninterrupted path in mid-air is what performing Koltès feels like. Leaving a sentence fragment hanging in mid-air for too long before reconnecting it to the main clause is impossible. Delivery requires speed, with speed comes risk; but Koltès has orchestrated the partitioning of the text so that it provides a series of hooks on which to hang the delivery in the form of rhetorical conjunctions or relative pronouns or alliterations or assonances. When performing the text in French, what aided Dalmasso in rehearsals was to visualize the different strata of fragments of sentences in front of him and then rewire them before they crashed to the floor, as though juggling with blocks, making sure that the pressure between the blocks was tight but loose enough to throw the blocks up in the air, catch them and rearrange them in a different manner. And while, as we have said, it was Stickman's vocal performance which interested us, it was also his physical performance – both of these convey a very particular rhythm. Clive Scott (1993, p.23) notes that 'Plato described "rhythm" as "the name for order in movement"' and that 'the kinetic element is crucial' and the physical, kinetic, element of Stickman's performances of our translations, in drumming and movement, were indeed a crucial building block in the construction of the rhythms of the final translation.

The retention in translation of Koltès' rhythmic structures, his musical score, enabled us as translators, in conjunction with Stickman as performer, to improvise like a jazz performer picking up on the rhythm of the text. Stickman's involvement facilitated this creative work on sound in particular. In translation we sought a new realization or variation. What Stickman brought to our translation was, in certain places, an identification of which rhythms were more successful, which rhymes gave him something to get his teeth into, where the text needed speeding up or slowing down, which sounds were most effective to perform. Our focus on retaining the complex syntax, word order, punctuation and avoiding expansion meant that our translation strategy was one which emphasized form over sense. However, this is not formal equivalence in Nida's sense as it included a distinctly dynamic element. It was a strategy which also emphasized sound over sense, or at least, which provided tones which coloured our interpretation, our musical variation. This strategy of sound over sense was pursued specifically to infuse the translated text with life and rhythm. A 'clear' translation with no particular sound features was always rejected in favour of a more opaque one rich in sound, or a straightforward, 'natural', translation rejected in favour of an unusual, awkward one, because we felt that what might be lost in meaning would be regained in rhythm. A good example of this is the phrase 'homme ou animal' in the French for which the most natural collocation in English is 'man or beast' but for which we employed 'man or animal'; it has assonance and alliteration which is there in the French but which 'man or beast' does not have while also providing us with an enforced pause because of the enunciation required in delivery, a pause which contributes to the rhythm of the piece. But this is not all, the deliberate similarity between humans and animals in Koltès' text is sustained more clearly with 'men or animals'. Example E is a perfect example of the way Koltès' text performs: teasing out the metaphor does not make the text more intelligible but recreating a rhythm, a system of sounds which performs the metaphor, does. Sounds provided us with the tones which overarch the whole piece, and soundscapes replaced the elusive alexandrine stanzas in our rhythmic architecture. Sound became our rhythmic scaffolding within which we built the text. The sound rhythm works to a large extent on rhymes, alliteration and assonance and can be seen in most of the examples we have used in this essay. An example of the importance of sound in alliteration, an example which was refined via Stickman's performance, occurs in example D:

mais ne me demandez pas de deviner votre désir [. . .] (2004, p.12)

but do not demand of me to divine your desire [. . .] (Baines and Dalmasso, unpaged)

As noted above, this is also an example of text having a performative function. The Dealer is constantly trying to stop the Client from leaving and his only weapon is language, so the delivery is that of a man in such a hurry to get his words out that he produces the sound of someone stuttering in haste. In order to reproduce this, we needed to ensure that the translation maintained the strong alliteration and pace of the original. We have also discussed above the following example as another illustration of the text performing the actions it describes, and as an example of Koltès' work with sound patterns. In translation, it works both as an example of the retention of the performative function of text but also as an example of the creative work we have done on maintaining or creating sound patterns:

Mon désir, s'il en est un, si je vous l'exprimais, brûlerait votre visage, vous ferait retirer les mains avec un cri, et vous vous enfuiriez dans l'obscurité comme un chien qui court si vite qu'on n'en aperçoit pas la queue. (2004, p.15)

My desire, were there one, were I to divulge it, would burn your face, have you retract your hands with a scream and you would flee into the dark like a dog who runs so fast his tail fades away. (Baines and Dalmasso, unpaged)

For the gradually accelerating melody of the French (see detailed description above, we have imported a similar shift of gear in the sentence at 'and you would flee into the dark like a dog which runs so fast his tail fades away'. The use of the inversion of subject and verb in 'were there' and 'were I' beforehand slow the first half of the sentence down. And while we do not have the tail rhythmically separated off, marked out in sound so as to draw attention to its disappearance as in the French, the sound of the English produces a different but similar effect and enables the tail to slip away. In addition, we have provided some alliteration in 'dog' and 'dark' and 'fast' and 'flee' which echoes that of 'aperçoit' and 'pas' and 'court' and 'queue'.

Our creative work on sound enabled us to attenuate the strangeness of the syntax which we had largely retained. This tightening of the syntax with deliberately creative work on sound enables both the performer and the audience to better deal with the stretched cohesion of the complex syntax. The repetition of sounds gives the performer and the audience a thread to follow alongside that of the syntax or alongside that of the meaning or of the plot. It is only in the final collision of the concurrent threads at the end of the play, in the final 'rien' ('none' in our translation) that the truth is revealed, nothing was said, nothing was heard:

G
Le Dealer
S'il vous plaît, dans le vacarme de la nuit, n'avez-vous rien dit que vous désiriez de moi, et que je n'aurais pas entendu?
Le Client
Je n'ai rien dit; je n'ai rien dit. Et vous, ne m'avez-vous rien, dans la nuit, dans l'obscurité si profonde qu'elle demande trop de temps pour qu'on s'y habitue, proposé, que je n'aie pas deviné?
Le Dealer
Rien.
Le Client
Alors, quelle arme? (2004, p.61)

The Dealer
Pray tell me, in the tumult of the night, did you not speak of one desire that I may not have heard?
The Client
I spoke of none; I spoke of none. And you, did you not offer one, in the night, in the obscurity so profound that only time can tame, that I did not divine?
The Dealer
None.
The Client
So, which weapon? (Baines and Dalmasso, unpaged)

Emphasizing the soundscapes instead of the syntax gave us the freedom to ensure that we adhered to our intention to retain the spaces between the text, between the metaphorical notes, to subtract rather than to add, with each sound calling for its corresponding silence, each beat calling a non-beat as in the syncopated rhythms of reggae.

Conclusion

This essay has provided an analysis of how specific elements in a text which invites one type of performance can be translated across languages into another quite specific type of performance using different elements which fulfil a similar function. The way in which the rhetorical conjunctions and relative pronouns follow the rhythm in the French cannot be recreated if a performable text is sought, that is, a text within which performability is defined as rhythm, so the emphasis had to be placed on sound and 'musical arrangements'. What we are tempted to call *percussive theatre* is Koltès' creation of a rhythmic architecture consisting of a combination of malleable versification which evokes alexandrines, or themes, and overarching complex syntax, or 'rhythm section', which build specific strata to construct breathing and pace of delivery, along with sonorities, or 'interpretations', in repetition, rhyme, alliteration and assonance. These features are transferred in translation from one performed text to another, with the main difference being the focus on sound and rhythm to compensate for the lack of transferability of the complex syntax. However, the English play text could not have been created in abstract conditions, the practical input of jazz poet and musician Stickman was crucial. We believe that the work we have done is an important step in demonstrating that the relatively rare practice of close collaboration between translator and director/actors can be made into a much more direct exchange than it often is. In general we believe a performer can benefit enormously from having to be confronted by the strangeness of the original text and by wrestling with its rhythm. In our practice, we ask the performer to be 'l'étranger qui ne connait pas la langue, ni les usages, ni ce qui ici est mal ou convenu, l'envers ou l'endroit, et qui agit comme ébloui [. . .] (Koltès, 2004, p.33), 'the stranger who knows neither lingo nor protocol, neither lore nor custom, inside nor out, and who acts as though bedazzled' (Baines and Dalmasso, unpaged).

Notes

1. (*Translation by Baines and Dalmasso – all translations in this essay, unless otherwise indicated, are by Baines and Dalmasso.*) La langue française, comme la culture française en général, ne m'intéresse que lorsqu'elle est altérée. Une langue française qui serait revue et corrigée, colonisée par une culture étrangère, aurait une dimension nouvelle et gagnerait en richesses expressives à la manière d'une statue antique à laquelle manquent la tête et les bras et qui tire sa beauté précisément de cette absence-là.
2. For example:

    (*6*) 'Si vous marchez dehors,
    (*12*) à cette heure et en ce lieu, c'est que
    (*12*) vous désirez quelque chose que vous n'avez pas
    (*12*) et cette chose, moi je peux vous la fournir'

    Strictly speaking this segment has nine syllables. However, it is the first line of the play and serves as an exposition. Since the only spatial and temporal references are 'cette heure' and 'ce lieu', it is conceivable (and indeed how Dalmasso performed it) that the Dealer would utter them slowly by stressing every syllable. Moreover, the assonance in 'dehors/heure' calls for an emphasis on the second word. Thus, the segment could be performed as a twelve syllable unit, with 'heure' pronounced as two syllables and with a diaeresis (pronunciation of two adjacent vowels in two separate syllables rather than as a diphthong) on 'lieu': 'à cette heure/ et en ce li-eu, c'est que'. See also extract of the beginning of *Dans la solitude des champs de coton* as rehearsed by Patrice Chéreau on audio CD with Benhamou (1996).
3. See Jacques Roubaud (2000).
4. See Jeffrey Wainwright (2004) for comparison.
5. Although there is not much evidence of performance or slam poetry influencing theatre, our experiment is not an isolated example of theatre using slam and Hip Hop culture, see in particular the work of the Theatre Royal Stratford East in London in their 2003 Hip Hop version of *A Comedy of Errors*

called *Da Boyz*, the street dance 2006 production of *Pied Piper* and 2007's production of Genet's *The Blacks*, for example.
6. Koltès travelled to Africa, specifically Nigeria, Mali, Senegal and the Ivory Coast and set the 1979 play *Combat de nègre et de chiens* in an African context. He also travelled to New York on a number of occasions and it was here that, according to Patrice Chéreau in the documentary mentioned above, the fleeting exchange with a dealer which is at the origin of *Solitude* occurred. It was in New York that he came across rap music as early as 1983 in Washington Square and it is his interest in 'black' music, in rap and reggae which is relevant here.

References

Benhamou, A.-F. (1996) *Koltès: Combats avec la scène*, special issue of *Théâtre d'aujourd'hui* (5) (includes audio CD of *Dans la solitude des champs de coton* rehearsed by Patrice Chéreau)
Berman, A. (2004) 'Translation and the Trials of the Foreign', in Venuti, L. (ed.), *The Translation Studies Reader* (London: Routledge), pp.284–97
Boeglin, B. (February 2001) 'Roberto Zucco est une pièce où le sentiment d'urgence est très fort' *Le Magazine Littéraire* (395) p.46
Chernoff, J. Miller (1980) *African Rhythm and African Sensibility: Aesthetics and Social Action in African Musical Idioms* (Chicago: University of Chicago Press)
Delgado, M., and D. Fancy (May 2001) 'The Theatre of Bernard-Marie Koltès and the "Other" Spaces of Translation', *New Theatre Quarterly* (XVII: 2), pp.141–60
Franzon, J. (2008) 'Choices in Song Translation', in S.-S. Sebnem (ed.) *Translation and Music, The Translator* (14: 2), pp.373–400
Hart, M. (with J. Stevens) (1990) *Drumming at the Edge of Magic: a Journey into the Spirit of Percussion* (New York: Harper Collins)
Koltès, B.-M. (1989) *Combat de nègre et de chiens* (Paris: Minuit)
Koltès, B.-M. (1999) *Une part de ma vie* (Paris: Minuit)
Koltès, B.-M. (2004) *Dans la solitude des champs de coton* (Paris: Minuit)
Meschonnic, H. (1982) *Critique du Rythme* (Paris: Verdier)
Rogers, A. (1994) 'Rhythm and the Performance of Organization', in P. Auslander (ed.) (2003) *Performance: Critical Concepts in Literary and Cultural Studies* (London: Routledge), pp.353–404
Roubaud, J. (2000) *La Vieillesse d'Alexandre* (Paris: La Découverte)
Scott, C. (1993) *Reading the Rhythm* (Oxford: Clarendon Press)
Wainwright, J. (2004) (trans.) 'In the Solitude of Cotton Fields', in D. Bradby and M. M. Delgado (eds.) *Bernard-Marie Koltès: Plays: 2* (London: Methuen), pp.187–215
Websters Dictionary (2009) (online) Available at: http://www.yourdictionary.com/rhythm (Accessed 20 January 2008)

4
The Theatre Sign Language Interpreter and the Competing Visual Narrative: the Translation and Interpretation of Theatrical Texts into British Sign Language

Siobhán Rocks

Throughout this essay I use the following terms: *Deaf* for pre-lingually deafened first language sign language users, *hearing* for non-deaf spoken language users, and, after theatre semiotician Keir Elam (2002), I have used the terms *dramatic text* for the scripted dialogue written *for* the theatre, and *theatrical text* for that produced *in* the theatre – the complete performance.

Deaf people are embedded in the dominant hearing culture, but it is essential to make clear that the Deaf community has a very distinct and separate cultural and linguistic identity of its own. Hearing people glean much of their world knowledge from hearing incidental chatter, background noise, radio, music, and so on. In contrast, Deaf people's incidental learning is through visual markers. This gives them a different world perspective, and in interpreting and translation terms we must treat the Deaf audience as 'foreign'.

The preferred first language of the British Deaf community is British Sign Language (BSL). BSL is not coded English, indeed no signed language is based on the mother tongue of its country. Sign Language is not international (although International Sign Language is an Esperanto-type signed language developed relatively recently and used mainly for international conferences). BSL is a non-linear spatial-visual language with linguistic properties very different from those of English. Sign language is not only made with the hands; for example, functions such as tone, mood, questions, counterfactuals and hypotheticals are all conveyed by facial expression, simultaneously expressed with, and modifying, the manual utterance.

Deaf people have historically been given poor access to education, identified by the dominant hearing community as 'disabled' rather than as members of a linguistic minority. They also have no particular culture of attending non-Deaf theatre because sign language interpreted theatre is a relatively recent development.

Indeed, sign language interpreting itself is a relatively young profession. The public perception is that the sign language interpreter is based firmly in the community and that, to a degree, is the case. The first sign language 'interpreters' were siblings or children of Deaf people, ministers or social workers. Today the majority of work undertaken by professional sign language interpreters is still community based, and BSL interpreters do not have the academic tradition of literary translators or spoken language conference interpreters. This situation is beginning to change, however. The University of Leeds Centre for Translation Studies has, since 2003, included BSL interpreters on its MA Interpreting Studies programme. This is a much-needed step forward, yet there is still no specific training for BSL interpreters working in highly specialized settings.

Thanks to the introduction of the Disability Discrimination Act, and the growing recognition that British Sign Language is the first language of Deaf communities throughout the UK, there has been a rapid and substantial increase in the number of mainstream theatres providing interpreted performances of their productions and, as a result, more Deaf people are being given the opportunity to experience theatre.

I made the following observations at a selection of BSL interpreted performances at different UK theatres during 2006:

i. During the performances, the interpreters attempted to sign all dialogue, leading to constant production of signed language and, consequently, the Deaf spectators were rarely given the opportunity to see any stage activity.

ii. The interpreters' signed renditions of the dialogue often lagged so far behind the spoken lines that it impacted on the accuracy of interpreter's role shift (due to the interpreter signing a number of lines behind the actors). Moreover, since interpreters were often still signing a previous speech during periods of no dialogue, the audience were again unable to look to the stage for salient visual information or, indeed, to identify which character was speaking, and at what time.

iii. The regular intrusion of source language, in the form of 'signed English' (BSL signs in English sentence structure) in the interpreter's rendition of the dialogue, and the literal interpretation of source language idioms and metaphors, resulted in the target language being, to varying extents, incomprehensible.

These observations suggested that the interpreters:

i. Were not taking into account the complex nature of the theatrical event, and the interdependent relationship between the spoken dialogue and the non-verbal and aesthetic elements of a production.
ii. Were employing the accepted general interpreting strategies in a highly specialized, complex and time-constrained setting.
iii. Did not understand sections of, or had not sufficiently prepared, the source text.

To explore these observations, in 2006 I conducted interviews with practising BSL theatre interpreters apropos of their approaches to the task of interpreting theatre. I discovered that the majority of sign language interpreters have no theatrical background. To some extent this is not surprising, since there is no specialized training in interpreting for theatre, and the BSL interpreting community reflects the general population in that regular theatregoers are in the minority. My research also found that a significant number of interpreters were applying general simultaneous interpreting approaches used in dialogue and conference settings, rather than recognizing the complexity that occurs with the theatrical setting. Of the BSL interpreters interviewed, sixty-four per cent accepted that when interpreting a theatrical performance there would be a naturally occurring 'lag time' (Cokely, 1992, p.42) between the spoken utterance in the original, and the delivery of the interpreted signed utterance. Of course, the theatre sign language interpreter is able, in most cases, to have access to the dramatic text weeks or months in advance of the theatrical event, and access to performances of the production in advance of the interpreter-mediated event, in order to prepare the interpretation fully. Interpreting scholar Mariachiara Russo (1995, p.344) describes this situation as: 'protected' in the sense that the interpreter has greater leeway to enhance his or her performance. However, I found that twenty-seven per cent of sign language interpreters spent only between two and five hours in preparation for the assignment, and a further twenty-seven per cent spent a maximum of 20 hours in preparation.

Clearly, the understanding of the spoken dialogue of the performance is essential for the audience to follow plot, identify themes, characters and so on. Due to the visual nature of sign language, the BSL interpreter is required to be visible to the whole audience throughout the performance, and the interpretation has to respond to what goes on onstage in the moment of performance. Each performance is also subject to spontaneous modifications, mistakes, inspiration, audience response, or any other circumstance prevailing at the time (Esslin, 1998, p.299). So, like the actor, the BSL interpreter has to be able to respond to any situation that may present itself (a requirement of simultaneous interpreting in any setting) whilst maintaining the 'reality' of the world of the play.

Although the activity of theatre sign language interpreting might appear to be a straightforward simultaneous interpreter-mediated event, it cannot be treated as such because the types of language and interactions we encounter in theatre are not the ones that occur in everyday interpreting situations. Dramatic dialogue is a *representation* of spontaneous discourse. Not only what the characters say, but also how they participate in the dialogue – their interactional patterns, turn taking and so forth – all are signifiers providing information for the audience. And there is an additional feature: dramatic dialogue is not meant for the interactants in the drama, but for a third party – the spectator. Bakhtin (1986, pp.61–2) differentiates this observed dialogue from everyday conversation by identifying it as *secondary speech genre* (where social dialogue meant for the speaker and addressee is *primary speech genre*). Dramatic dialogue lacks redundancy, is layered with meaning, and often has 'a tight dramatic structure [. . .] consisting of an Exposition, Development, Climax and Denouement [. . .] it is purpose-driven' (Remael, 2004, pp.107–11); it functions retroactively by reinforcing preceding events and proactively by moving the plot forward.

The fact that BSL has no written form means that a written translation in the traditional sense cannot be developed. Yet we can enter into a 'translation' process where, like the translator of written text, we make informed decisions about how to render in the target language and compensate for specific sections of the dramatic text, be they plot-carrying or character-defining elements, intertextual references, wordplay, issues of cultural distance, and so on, well in advance of the interpreting event.

That said, until we see how the text will be performed, although we may assume to understand the *dramatic* text, we cannot know how it will be transformed by the makers of the *theatrical* text.

For the audience, the opportunity to 'read' the constantly changing stage picture is of vital importance to their understanding of the piece as a whole. Although the hearing audience member can choose to look at any part of the stage at any point during the performance, irrespective of any ongoing dialogue, the Deaf spectator cannot. For the Deaf audience, restricted to visual channels of communication only, the presence of the BSL interpreter means that the plurality of those visual channels is increased, and, since the vast majority of signed performances are interpreted from the side of the stage, that forces the Deaf audience to negotiate *two* competing points of focus: the signed dialogue of the interpretation, and the continuing visual narrative of the performance.

BSL interpreters therefore cannot confine themselves to the two extremes of source language and target language because there is an additional issue: the negotiation of the plurality of simultaneously occurring visual information that supports, enhances and may play against the delivery of the dialogue. It is essential then, that the interpreter is familiar with the performance to be interpreted, either through access to latter-stage rehearsals and/or an audio-visual recording of an early performance of the production. This forces the interpreter to make decisions based on the interactions of all three areas – translation, interpretation and the performance (as shown in Figure 4.1), engaging with the piece from the beginning of the 'translation' process to the end of the interpreted performance (Gambier, 2003).

In the areas of media (film and television) interpreting and translation (dubbing and subtitling), research and discussion is already underway in an attempt to develop a specific methodology of 'screen translation',

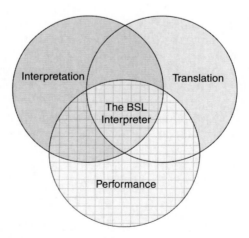

Figure 4.1: The interaction of the three areas concerning the theatre sign language interpreter

which encompasses the multimodality of audiovisual communication. We can see the similarities of genre in Russo (1995, p.343):

> For the media [. . .] perhaps more than in any other field of translation, the dynamic, semantic and pragmatic equivalence between the [source language] and the [target language] is not enough because also the aesthetic impact plays a fundamental role.

And in Josélia Neves (2004, p.135):

> In audiovisual translation the problems which arise are somewhat similar to those of literary translation with the extra stress that the fidelity factor is dictated by constraints that lie beyond words or languages [. . .] in audiovisual translation fidelity is particularly due to an audience that, like the receiver of the simultaneous interpretation, is in need of communicative effectiveness, rather than in search of artistic effect [. . .] or of exact equivalence . . .

Like the audiovisual translation, the BSL rendition of the spoken dialogue has to be presented concurrently with the performance event, so that any translation developed without reference to the theatrical text will be informationally incomplete. The signed rendition is bound to the temporal restrictions of the performance, the live interpretation begins and ends with the performance, and the signed rendition of dialogue must be timed to correspond to the utterances of the actors, yet there must be time allowed for the audience to witness salient stage activity.

The minimum the theatregoer can expect when watching a play is to understand the plot and identify the characters. It is a fundamental requirement then, for the foreign audience's understanding of the dramatic text, that all essential plot-carrying and character-defining elements are translated but, in theatrical texts, those factors are not expressed exclusively through dialogue.

Observation of BSL interpreters in the field suggested that they were not aware of the complex nature of the theatrical event. The following examples were observed at a BSL interpreted performance of Ibsen's *Hedda Gabler* in 2006.

As Hedda's husband, Tesman, sees his aunt to the door (offstage), he is heard saying goodbye and thanking her for the slippers she has brought him. Simultaneously, we see Hedda furiously pacing the room, with clenched fists, finally flinging back the curtains at the window and looking out impatiently, until Tesman's return. In this instance the BSL interpreter chose to sign the offstage dialogue, drawing the audience's attention away from the activity onstage. Once the audience has missed a visual element so significant that its informational value is far greater than that of the spoken text, the meaning is lost, and unrecoverable.

Similarly, later in the piece, after Judge Brack's exit, Hedda's rummaging through her husband's writing desk was effectively upstaged by the interpreter signing the maid's voice heard from offstage, seeing the Judge to the door. Since the audience had already seen both Tesman's aunt and the Judge exit the stage, each having said 'goodbye' to Hedda, in each case the offstage dialogue had no plot-carrying function and should not have been translated at the expense of essential visual plot-carrying and character-defining information.

In order to ensure that vital visual information is not lost, it is often necessary to manipulate the timing of the delivery of sections of the BSL rendition. This allows the shift of the audience's focus, in advance of the salient stage activity, from the interpreter to the stage in order that the spectator can maintain, as far as possible, the actor–audience relationship and their engagement with the performance. Once the audience's attention is focused on the stage, in order to return the attention to the signed dialogue, it is perfectly legitimate for the interpreter to add an appropriate linguistic element at the beginning of a character's utterance, a device that functions as a signal for the audience to return their attention to the interpreter.

The next example was observed at BSL interpreted performance of Shakespeare's *Richard III* in 2006. In this production there was a direct and clear relationship between Richard and the audience beyond the scripted monologues and asides. At various points during the performance, Richard would look directly at the audience with a wink, a smirk or a nod, openly commenting on the action or dialogue taking place, a device reinforcing a much more intimate and conspiratorial relationship with the spectator.

Unfortunately in this case, the BSL interpreter's rendition consistently lagged behind the spoken dialogue, there was no flexibility in the timing of the signed dialogue, and as a result the Deaf spectator was unable to share in this intimacy of actor–audience relationship.

Taking into account the plurality of simultaneously occurring aural and visual information, the BSL theatre interpreter's source text must be the performance itself. This essay advances a model approach to sign language interpreting, specifically for this complex assignment (see Figure 4.2), since there is no established model of translation or interpretation

Performance
Translation issues cannot be confined to only source language and target language. The BSL interpretation is bound to the essential visual elements and temporal constraints of the performance. BSL interpreting for the theatre is unique because the source text is the theatrical text, the complete performance.

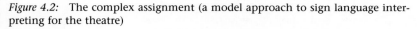
The complex assignment

Translation
We must fulfil the requirements of fidelity to sense, 'spirit', communicative effectiveness, and the production of a similar response, translating essential plot-carrying and character-defining elements. Yet the translation must be flexible in order to respond to circumstances prevailing during the performance.

Simultaneous interpretation
The BSL interpreter interprets the dialogue simultaneously into BSL. However, the dramatic text has a tight structure, is purpose driven, layered in meaning and lacks redundancy. The actor–audience relationship is an oblique one. The interpretation is temporally bound to the performance.

Figure 4.2: The complex assignment (a model approach to sign language interpreting for the theatre)

that fits this unique setting. The model has been developed through my own practice and that of my colleague Alex McDonald, and is a multidisciplinary approach to preparing dramatic and theatrical texts for BSL interpreted theatrical performance. Of course, as with any other translation, the theatre sign language interpreter must fulfil Nida's basic requirements of a translation (1964, p.164):

- making sense
- conveying the spirit and manner of the original
- having a natural and easy form of expression
- producing a similar response.

The dramatic text contains plot-carrying and character-defining elements essential to the understanding of the work. However, this text is enhanced in the theatrical process by the non-linguistic and aesthetic factors that also develop plot and define character. The model assumes that the performance is the BSL interpreter's source text, and that it is necessary to examine performance in order to:

- identify linguistic plot-carrying and character-defining elements as a priority;

- identify those sections in which the visual element is of greater informational value than the spoken text;
- identify points at which the timing of the signed rendition of dialogue needs to be manipulated, in order that the Deaf audience are able to witness visual elements essential to the comprehension of the piece;
- identify the qualities of the various and distinct performances. Since the Deaf spectator cannot look at the actors when he or she is watching the signed dialogue, those qualities must be reflected in the interpreter's rendition so that the characters represented are not only identifiable, but also 'come across as having slightly more complex personalities than just being goodies or baddies' (Herbst, 1995, p.269).

In this setting, we cannot isolate the spoken dialogue from the polysemiotic web of the theatrical text. It would serve little purpose then, to attempt to translate at sentence level, since it is the whole scene, and its relationship to other scenes within the overall arc of the piece, that carries the complete intended meaning. We have to assume, then, that the scene is our smallest unit of translation (Herbst, 1995, p.270).

Decisions about how to render in the target language specific sections of the text cannot be made in the moment of performance, and need to be fully prepared in advance, for the reasons adduced earlier. Yet, throughout a performance interpreters must make themselves as available and adaptable as the actors. The fact that a performance is live means that it is subject to the effects of circumstances beyond the control of any participant in the performance. Since the BSL rendition of the spoken dialogue is temporally and spatially bound to that performance, the sign language interpreter must develop what theatre practitioners term *élan*, a readiness to respond to whatever circumstance presents itself. The actor's *élan* – often mistaken for true spontaneity – does not come from a sense of riskiness or a lack of preparation; it is based on a depth of understanding of the piece that allows the actor to deviate, if necessary, from the text and yet still retain fidelity to both the character and the work.

Just as the actor, when speaking his or her dialogue, is required to make that rehearsed dialogue appear as if it is being uttered for the first time in a moment of spontaneity, the interpreter must also reflect that apparent spontaneity in their signed rendition.

The following is an example of the use of the model, taking the theatrical text as source text. The production of *Sing Yer Heart Out for the Lads*, written by Roy Williams, was directed by Marcus Romer for Pilot Theatre, and was interpreted by Alex McDonald in 2006. *Sing Yer Heart Out for the Lads* asks what it means to be British in the twenty-first century, and confronts issues of race and nationalism.

The scene lasts for approximately one minute. The location is inside a pub. The stage is set on three levels: the higher upstage level with the bar centre back, and pub doors to stage right; the mid-stage area, a shallow landing two steps lower; a further two steps down, the main downstage area, set with a pool table downstage right, table football stage left, and beyond further left, a door signed 'Gents'. There are various pub stools downstage centre. Four (white) members of the pub football team, dressed in England away shirts, are downstage, with their friends Alan and Lawrie (dressed in regular jeans and shirts). Mark (black, early thirties), the elder brother of Barry, is asking Alan (white, late forties) why he gave all the pub football team members except Barry a lift from the match back to the pub.

At this point Barry, dressed in an England away shirt with a flag of St George painted on each cheek, bursts through the pub doors and dances to downstage centre, body-popping, moonwalking and singing in front of the group. Throughout Barry's dancing, his team-mates sing a verse and chorus of the 1990 England World Cup song, New Order's 'World in Motion'. During the celebrations, Lawrie remains apart, observing from the pool table. Mark, Barry's brother remains on the upstage level, stage right of the bar, observing the scene with contempt. Alan observes the celebrations at first from the higher level, stage left of the bar. As the singing ends, Barry falls to his knees, arms outstretched, to allow his team-mates to rub his head for luck, as if he is their mascot. As the group disperses, Barry shouts 'Anyone else?' (Williams, 2002, p.35) as Alan descends to the downstage area, steadies Barry's head with one hand, rubs it with the other, and moves away, rubbing his fingers together. Barry finally stands and turns, notices Mark and addresses him with 'Alright, bruv?' Mark responds, 'Can I talk to you?' (p.36) nodding towards the door of the Gents' toilet.

If the interpreter, in this case, chooses to concern himself solely with the dramatic text, then the Deaf audience will simply see the signed rendition of the song lyrics, followed by the line 'Anyone else?' and so on. However, if we take the scene as *performed* as the source text, we find a great deal more information.

This scene occurs early in Act One, and provides important plot, character and relationship-defining markers. It is vital that the dialogue between Mark and Alan, immediately preceding Barry's entrance, be interpreted since it provides the essential information that Barry, the only black team member, was also the only one not to be given a lift from the match back to the pub. Barry's kneeling marks his willingness to play the 'Uncle Tom' role, that of the subservient black man compliant with the white majority. The flags painted on his cheeks show that he is prepared to go further to gain acceptance into the group. The head rubbing marks Barry's lower position within the group, reinforcing his brother's suspicion that he is treated in this way because he is black. Other elements such as the costuming of the characters, and the positioning of each character in the scene, provides a visual reinforcement of their relationships with each other and their attitude to the activity during the scene. That Alan comes over to rub Barry's head, his manner, and that he does it separately from the football team, is a marker of two things: firstly that he wants to be seen as one of them, and secondly, his independence from them. The reason for this becomes clear later in the play when we discover that Alan is, in fact, a political activist for the far right, looking to recruit members whilst maintaining a distance from any involvement in racial violence. He has a very brief private discussion with Lawrie immediately after this scene, informing him that he has been accepted as a member of his un-named 'political' party.

We have to remember that the Deaf audience do not see the stage in the same way as the hearing audience, because they have to use the sign language interpreter. The interpreter's decision in this case, however, does not simply concern whether the Deaf audience will find the interpreting of the song too 'foreign' an experience, or if the stage picture is more 'interesting'. The fact remains that only a thorough examination of the scene in its context, will allow the interpreter to decide whether it is the spoken or visual element that carries more meaning for the target audience.

In this case, the interpreter chose to allow the stage to speak for itself. The dialogue between Mark and Alan, immediately before Barry's entrance, was interpreted, yet the timing of the signed utterance was adjusted so that it finished a moment before the spoken dialogue. The Deaf audience's attention was directed to the stage by the interpreter giving focus to the stage, in enough time for them to witness Barry's entrance. The following scene was allowed to play with no input from the interpreter, until Barry stands, notices his brother and says 'Alright, bruv?'

At this point the interpreter drew the attention of the audience away from the stage by adding a sign that can be glossed as 'come over here' to the beginning of the signed rendition of Mark's line 'Can I talk to you?'

Here the additional, yet appropriate linguistic element functioned as a device to bring the audience's attention back to the interpreter without altering the intent of Mark's utterance, also compensating for the nod of Mark's head indicating the toilet.

This required level of depth and detail would seem reasonable to the translator of a dramatic text and the theatre company commissioning a written text. What makes the theatre sign language interpretation unique is that this level of preparation is necessary for both the dramatic and theatrical texts.

Like the translator, the sign language interpreter should be qualified to make these decisions about the perceived intent of the playwright but also, to an extent, that of theatre makers. However, the research indicates that this, in the main, is not the case. Part of the problem is that very often the sign language interpreter is not theatrically competent. The situation is exacerbated by the fact that the theatre practitioner is almost never involved in the staging of the interpreter-mediated event. Responses from sign language interpreters reflected my own experience:

- the overwhelming majority of interpreters cited their point of contact as the administration department of the theatre, not the theatre practitioner;
- the BSL theatre interpreter is usually hired by virtue of availability, not by capacity to do the work;
- in the case of visiting companies, the interpreter is hired to interpret the show *after* the production is on the stage;
- even when the production is in-house, the interpreter is not invited to rehearsals, and is regularly refused access to rehearsals even if requested;
- the theatre interpreter translates and rehearses alone;
- the interpreter rarely gets contact with the makers of the piece until the night of the performance to be interpreted and, since the interpreted performance is usually a one-off, the first opportunity for the interpreter to work with the actors is in front of the paying audience.

The research found that almost seventy-five per cent of BSL theatre interpreters interviewed felt it necessary to have a dialogue with the makers of theatre in order to improve the quality of their work, however only nine per cent had been able to achieve that dialogue. The fact that a member of the theatre's administration department might initially engage a BSL interpreter presents no real problem if there are clear channels of communication to a member of the company, or indeed, awareness that those levels of communication may be necessary. It appears that presently, however, the interpreters' point of contact at the theatre is usually an administrative not an artistic one, and there is a clear discrepancy between what the interpreters feel they need in order to do a better piece of work, and what is generally available to them.

This lack of dialogue between interpreter and theatre practitioner serves to reinforce the misplaced assumption that BSL interpreted theatre is simply a matter of interpreting a prepared text from the side of the stage, not dissimilar to the conference interpreting setting. The notion that the interpreter-mediated event is simply a mandatory provision for a disabled minority, and separate from the actual staging of a production, must be addressed.

We need to ask why the theatre practitioner is not engaged with sign language interpretation of his or her work, when the interpreter appears on stage, in the full view of the audience, throughout the performance. Drama is written and theatre staged with the intent of having a particular effect upon the spectators who witness it. That effect can be significantly altered by the visible presence of a sign language interpreter, affecting the stage picture for the whole audience.

The following example is from the BSL interpreted performance of *Hedda Gabler*. On stage, Judge Brack and Hedda sat in silence, staring at each other, for twenty seconds or more, the situation speaking for itself, allowing the audience to infer a great deal about their past and present relationship. In the middle of this intense quiet and stillness, the interpreter looked to the stage, then back towards the audience and signed 'Silence', effectively upstaging the actors and stealing the focus of the entire audience.

Had the director of this production been involved with the staging of the interpreted performance, it is no doubt safe to assume that he would have directed the interpreter to do nothing during this particularly significant scene.

It is not only inappropriate interpreting decisions that can cause a distraction. The functional approach to the interpretation of theatre, and lack of contact with the theatre practitioner, appears to have led to a type of 'default' response in which the interpreter, usually dressed in black, is lit in a fixed spotlight throughout the performance, including blackouts and scene changes. This forces the audience to work much harder to suspend their disbelief, particularly if time has moved on in the world of the play – the constantly visible interpreter anchors the spectator to real time, whilst the time in the world of the play is progressing.

As previously discussed, BSL interpreters have, so far, no access to training in highly specialized areas of interpreting and, as a direct result, have no guidance as to what might be required of them when entering the world of theatre interpreting. Neither does it appear that there is any significant channel of discourse or artistic support from the venues that employ the interpreters for their performances. This communication between interpreter and theatre practitioner is unlikely to develop if neither party has sufficient understanding of the other's area of expertise to engage in a meaningful dialogue.

This essay does not aim to be a criticism of theatre sign language interpreters, but of particular approaches that have developed as a result of circumstances prevailing in the field. The adherence to the model, and the acknowledgement of the performance as the source text, forces a shift in focus from the functional notion of access, to the artistic notion of creating theatre for diverse audiences. Like literary translation, the BSL theatre interpretation must be treated with rigour, and interpreters trained accordingly. It becomes incumbent on the makers of theatre to treat the interpreter-mediated performance with the same artistic integrity as any other performance. Both the theatre practitioner and the BSL interpreter must be aware that the inclusion of an interpreter establishes two competing points of focus: the signed dialogue and the visual narrative – both essential for the Deaf spectator's understanding of the performance. The initiation of a dialogue that engages theatre practitioners and informs them of the notion of good quality interpreted theatre, may enable us to re-define what is understood by, and to develop a shared language of 'sign language interpreted theatre'. However, it also means that sign language interpreters can only engage in that dialogue if they have a complete understanding of the nature of their task. This understanding can be achieved by the specialist training of sign language interpreters in the negotiation of the linguistic and visual complexity of the theatrical event, using the model advanced in this essay as a multidisciplinary approach to the BSL interpretation of theatrical texts.

References

Bakhtin, M. M. (1986) *Speech Genres and Other Late Essays*, V. W. Mcgee (trans.) (Austin: University of Texas Press)

Cokely, D. (1992) 'The Effects of Lag Time on Interpreter Errors', in *Sign Language Interpreters and Interpreting* (Burtonsville: Linstok Press), pp.36–69

Elam, K. (2002) *The Semiotics of Theatre and Drama* (London: Routledge)

Esslin, M. (1998) 'The Signs of Stage and Screen', in G. W. Brandt, (ed.) *Modern Theories of Drama: a Selection of Writings on Drama and Theatre, 1840–1990* (Oxford: Clarendon Press), pp.299–307

Gambier, Y. (2003) 'Screen Transadaptation: Perception and Reception', *The Translator* (9: 2), 171–89

Herbst, T. (1995) 'People Don't Talk in Sentences: Dubbing and the Idiom Principle', *Fédération Internationale des Traducteurs (FIT) Newsletter* (XIV: 3–4), pp.257–71

Neves, J. (2004) 'Language Awareness Through Training in Subtitling', in P. Orero (ed.) *Topics in Audiovisual Translation* (Philadelphia: John Benjamins), pp.127–40

Nida, E. A. (1964) *Toward a Science of Translating with Special Reference to Principles and Procedures Involved in Bible Translating* (Leiden: E. J. Brill)

Remael, A. (2004) 'A Place for Film Dialogue Analysis in Subtitling Courses', in P. Orero (ed.) *Topics in Audiovisual Translation* (Philadelphia: John Benjamins), pp.123–26

Russo, M. (1995) 'Media Interpreting: Variables and Strategies', in *Fédération Internationale des Traducteurs (FIT) Newsletter* (XIV: 3–4), 343–9

Williams, R. (2002) *Sing Yer Heart Out for the Lads* (London: Methuen)

5
Inferential Meaning in Drama Translation: the Role of Implicature in the Staging Process of Anouilh's *Antigone*

Alain J. E. Wolf

Introduction

The notion of inference, and, more specifically, Paul Grice's[1] theory of conversational implicature,[2] is relevant to the staging of translated drama on many counts. Playwrights generate inferences in their dialogues, and as translation scholar Kirsten Malmkjaer (2005, p.150) observes, the transfer of such inferences to contexts across languages is not unproblematic. The presence or absence of inferences in translated texts is frequently discussed (Baker 1992; Blakemore 1992; House, 2006; Malmkjaer 2005) as is the notion of 'context' which is seen to be a necessary component of cross-linguistic understanding. The most well-known account of what inferencing amounts to is Grice's (1975) model of utterance understanding. This essay will discuss the theory of implicature and inference and apply that theory to the staging in translation of a production of French playwright Jean Anouilh's 1944 play *Antigone* in which I performed.[3] I shall look at specific instances when the staging process of the play either helped towards the understanding of inferences or added inferences which were not there to begin with.

The Gricean account is mainly concerned with spoken utterances in what J. Lyons (1977, p.638) refers to as the 'canonical situation of utterance', for example 'one-one, or one-many [. . .] with all the participants present in the same actual situation able to perceive one another [. . .] and each assuming the role of receiver and sender in turn'. So it cannot be assumed that an account of inferencing along the lines of a Gricean model of utterance understanding will be readily applicable to the study of translated texts since they are produced in a non-canonical situation[4] of utterance subject to indeterminacy and ambiguity.[5] The next two sections are devoted to outlining Grice's (1975) theory of conversational implicature and its application to translated texts in general. I shall then address the issue of how a theory of implicature can explain translational phenomena and the staging process of Anouilh's *Antigone*.

Logic and conversation

In a paper entitled 'Logic and Conversation', Grice (1975) develops the notion of implicature starting with a distinction between a formalist and an informalist understanding of language. The formalist position concerns itself with the meaning of the formal devices of logicians, i.e. 'not', 'and', 'or', 'if', 'all', 'some', 'the', and constructs a system of valid inferences involving such devices. According to the informalist perspective, however, an expression in natural language can be guaranteed validity even if many of its inferences are not amenable to the treatment of logic (Grice, 1975, p.43). In an attempt to address these divergent views, Grice's strategy is to hold on to the meaning of the logical formal devices and to allocate the effects associated with their natural language counterparts to the category of implicatum, that is, whatever is implied beyond what is merely said.[6]

An example taken from Grice (1975, p.44) will serve to illustrate this strategy: Consider the formal device '→' (if). In logic, a sentence of the form 'P → Q' (if P then Q) could only be considered to be false in cases where P is true and Q is false. This is because a false conclusion cannot be derived from a true premise. The sentence would still be classified as true in logic if P is false. Let us translate this into natural language, using Grice's (1975, p.44) example:

(A) He is an Englishman, he is, therefore, brave.

We see that, in logic, the sentence would be true even if the statement that 'he is an Englishman' is not true. This would appear to go against the grain of our conventional natural language assumption that what is reported in the second member of an 'if-sentence' is a direct consequence of what is stated in the first. In Grice's view, however, the real meaning of 'if-then' remains the logical one of derivation of a true conclusion from a true premise. The notion of a consequential link between first and second clause is what he refers to as a *conventional implicature* which is not part of the real meaning of the words.

The implicatures Grice (1975, p.45) writes about in most detail are a 'subclass of non-conventional implicatures', i.e. conversational implicatures. Given that exchanges are recognized by participants to be rational and have 'a mutually accepted direction' (Grice, 1975, p.45), the general Co-operative Principle (hereafter, CP) will normally be observed by speakers and hearers:

> Make your conversational contribution such as is required, at the stage at which it occurs, by the accepted purpose or direction of the talk exchange in which you are engaged. (1975, p.45)

A Kantian framework of four categories follows:

> Maxims of quantity (how much/little information)
> Try to make your contribution as informative as is required: for the current purpose of the exchange.
> Do not make your contribution more informative than is required.
> Maxim of quality
> Try to make your contribution one that is true, specifically:
> a) do not say what you believe to be false
> b) do not say that for which you lack adequate evidence.
> Maxim of relation
> Be relevant.
> Maxim of manner
> a) avoid obscurity of expression
> b) avoid ambiguity
> c) be brief
> d) be orderly.

There are various ways in which a speaker might choose to disobey, or in Grice's terminology, flout a maxim:[7]

(1) by opting out (by refusing to talk about x);
(2) by violating one maxim because of a clash with another maxim, that is one cannot be as informative as one would like to be if one lacks the necessary information. In this case, the speaker will either flout quality or quantity;
(3) by blatantly flouting a maxim.

The deliberate flouting of the maxim of quality is particularly relevant in the context of works of literature as it accounts for instances of irony and metaphor. A speaker, 'A', talking about his close friend x who has betrayed a secret says to an audience:

> (B) 'x is a fine friend'.

It is obvious to 'A' and his audience that 'A' has said something which he does not believe to be true and the audience knows that 'A' knows that this is obvious to the audience. So 'A' must be trying to generate a conversational implicature of the sort: I am trying to convey the opposite of what I am saying.

Lastly, Grice (1975, p.57) lists five characteristics of conversational implicatures. They are:

(1) Cancellable, that is a speaker can explicitly choose to cancel the implicature generated by adding a clause, for example. John has three cows, and maybe more (example from Levinson, 1983, p.115).
(2) Non-detachable: since conversational implicatures are not part of the meaning of words, different types of expression can convey the same implicature. For example an ironic reading of 'John's a genius' with the implicature 'John's an idiot' can be conveyed by other expressions such as 'John is a mental prodigy, John's an enormous intellect' (examples taken from Levinson, 1983, p.117).
(3) Not part of the meaning of the conventional force of expressions to which they attach. For example the utterance 'John's stopped smoking' can generate the implicature 'you should stop as well' which is not part of the meaning 'John stopped smoking'.
(4) Not carried by what is said but by the saying of it, that is what is said may be true-what is implied may be false. For example a speaker might say (a) 'Herb hit Sally' implying misleadingly (b) 'He didn't kill her' where (a) is true and (b) is false (see Levinson, 1983, p.117)
(5) Indeterminate: an indeterminate array of implicata can be generated by an utterance. For example 'Keats's ink is pale' which could convey that his writing is dull, that it is difficult to read, that it is effete, or any of these and others, too.

Context, co-text and translated texts

Interest in the notion of context has grown since the early 1980s as linguists began to account for successful text comprehension in terms of an interaction between linguistic and non-linguistic knowledge.

Such a view of context raises interesting questions related to the comprehension of translated texts. Individuals communicating within the same linguistic and cultural communities are likely, in most circumstances, to share a greater number of beliefs and assumptions about the world than those communicating across such boundaries.

The consequence for communicators is that information which can be left implicit in certain intra-cultural situations may have to be made more explicit when used cross-culturally and cross-linguistically. Diana Blakemore's (1992, p.173) example, below, illustrates the extent to which a reader/hearer may fail to recover meaning:

(C) The river had been dry for some time. Everyone attended the funeral. (from Blakemore, 1992, p.173)

A reader/hearer would, in a significant sense, have failed to understand the text, if the cultural implications of the text were left implicit, namely, that if a river dries up, a river spirit has died, and that if this is so, a funeral takes place. It is in this sense that the text needs to be re-contextualized. Naturally, Sissala speakers could quite happily have had the conversation in (C) without being explicit about the context of what they said. This is because we assume a great deal of information to be shared between speakers/writers and hearers/readers. Indeed, an explanation of the kind I provided would be as redundant for Sissala speakers as would the explanation to a Westerner that a MacDonald's restaurant is a place where meat patties are sold between two pieces of bread (see Blakemore, 1992). Communicators, then, have to balance the risks of misunderstanding through implicitness against the risks of irrelevancy by being over-explicit.

We have, so far, addressed the notion of context only to show how accessing it across cultural boundaries might require special attention to the potential problems caused by lack of shared knowledge.

Let us now turn to the criticism of the pragmatic approach put forward by translation scholar Juliane House (2006, p.342 ff.) who claims that such a model is not applicable to written language and translated texts, in particular.

The pragmatic model, she argues, is exclusively concerned to account for the mechanisms of utterance understanding[8] in a dynamic context, constructed on-line between co-present speakers and hearers interacting over turns at talk in a canonical situation of utterance. By contrast, context in translated texts is static and the translator can only achieve 're-contextualisation' by 'linking the text to both its old and new connections' (House, 2006, p.343).

House's (2006) objections seem vague and I think it is possible to show that they are not wholly justified.

Granted there are difficulties in applying a pragmatic approach to the study of translated texts. Malmkjaer (1998) justly points out that

Grice's theorizing may not be a satisfactory theory of translation on the grounds that the meaning of S utterance depends on A's recognition of S's intention and that 'text severs that connection'.

In addition, wherever the inferences generated do not transfer to the translated text, we would have to argue, on Grice's account, that the original text has only been partially translated. The problem, Malmkjaer (1998) concludes, is that readers, not having perceived the inferences, will still judge the text-extract to be as an instance of translation. This is precisely the paradox encountered by source-text oriented theories (see Toury, 1998) when they attempt to account for translated texts which, on their own account, are non-translations. Clearly, these are problems which a Gricean approach to translation would also encounter.

But precise equivalence of the original author's intentions and effects, however, is not what is necessarily intended by translators. Inferences, which may be lost in places, may well be regained elsewhere in a fictional work, especially as the translator considers its entire co-text.

In this respect, it is important to emphasize the point that the connection between writers and translators is not as irretrievably severed as some would have us believe (House 2006).

Writers frequently resort to specific rhetorical strategies in order to clarify obscure references in the co-text of fictional works. In theatrical discourse,[9] for example, playwrights resort to an introductory scene in which a character informs the audience about what has just happened (see the role of the Prologue in *Antigone*) or to an informer/messenger (see the messenger who relates Antigone's and Haemon's death to the Queen).

In addition, as Malmkjaer (1998) points out, a great number of references are accessed by readers of texts and translated texts alike on the basis of their shared knowledge of entities in the physical world such as the sea, the sun, the moon, and so on.

Lastly, as we shall see in the next section, actors and directors gain access to a wide range of inferences, which might otherwise have been lost, through the staging process of a play. Means of working closely with text producers have increased through electronic communication and translators working in collaboration with playwrights and directors so that the connection between them is likely to be one of co-construction through the staging process.

In summary, although a Gricean account does not account for the difficulties outlined above, it is possible to argue (Malmkjaer, 1998, p.37), that the Gricean theory of conversational implicatures has been shown to give a successful enough account of non-literal meaning and that much in literary translation can be explained by the theory.

Examples

I shall begin this section by attempting to demonstrate how the Gricean theory of implicature can explain a great deal of what happens when translating with particular reference to Anouilh's play *Antigone*. The literature on implicature and translation (Baker 1992; Gonzalez 2006; Gutt 1991; Malmkjaer 1998; Mason 2006; Setton 2006) has found that inferences in the original text are often not generated by the translation when, for example, the interpreter does not recognize the pragmatic force of certain forms (Gonzalez 2006), when divergent contexts emerge between participants (Mason 2006), and when the literal meaning has been translated but the inferences have not (Malmkjaer 1998). This study was no exception and I shall begin this section by trying to determine what kinds of inferences generated by the source text did not transfer to the target text.

It has also been observed (Malmkjaer, 1998, p.37) that adjustments may be used by translators, e.g. adjusted punctuation, as compensatory strategies for the lack of inference transfer. As I have already observed (see introductory paragraph), I draw attention here to examples when the staging of the play contributed towards the recovery of inferences and sometimes even added inferences which were not in the original text.

Instances of inferences not recovered by the staging process include implicatures standardly derived from the conventional meaning of words. Consider the following text extract taken from Barbara Bray's translation (2000, p.12) of Anouilh's (1946) *Antigone* in which Ismene describes her fear of death at the hands of her uncle's militia:

> The guards will be waiting there, with their stupid faces all red from their stiff collars, their great clean hypocrites' hands, their loutish stare.

It is fairly clear that any English reader of the above would have no problem with making sense of the text as presented. Indeed, the staging process did not find it necessary to alter the text in any way. The above translation is a reasonably literal rendering of the following (Anouilh, 1946, p.27):

> Et là il y aura les gardes avec leurs têtes d'imbéciles, congestionnées sur leurs cols raides, leurs grosses mains lavées, leur regard de boeuf.

Compare the French phrase 'leurs grosses mains lavées' and the equivalent in the translation 'their great clean hypocrites' hands'. With the exception of the addition of 'hypocrites', which would, on the face of it, seem unjustified, the two clauses are an exact match: leurs = their, grosses = great, mains = hands, clean = lavées. The French phrase, 'leurs grosses mains lavées' however, standardly implicates the act of washing one's hands of something. From this phrase, a French audience can get in the implicature that the guards are likened to Pontius Pilate in that they abnegate responsibility. The translation attempts to make an adjustment by adding the word 'hypocrite', which leads one to think that the translator had recovered the implicature, but was not able to convey it successfully. Indeed, on first reading, it is unclear how Ismene could think that the guards whom she describes as louts are so sophisticated as to be hypocrites.

Consider too, the Prologue's description of Creon's guards (Anouilh, 2000, p.5):

> Lastly, those three red-faced fellows playing cards, with their caps pushed back on their heads – they're the guards. Not bad chaps. They've got wives . . . children . . . little worries the same as everyone else. But before long they'll be collaring the accused without turning a hair.

Again, any competent reader of English would have no difficulty with what is a communicative equivalent of the following (Anouilh, 1946, p.12):

> Enfin les trois hommes rougeauds qui jouent aux cartes, leur chapeau sur la nuque, ce sont les gardes. Ce ne sont pas de mauvais bougres, ils ont des femmes, des enfants et des petits ennuis comme tout le monde, mais ils vous empoigneront les accusés le plus tranquillement du monde tout à l'heure.

Aside from a number of strikingly dissonant translation choices such as the use of 'chaps' with its upper-class connotations for 'bougres' and 'red-faced' which tends to imply shame rather than drunkenness, let us first focus our attention on the use of the word 'tranquille' as this will give us a basis on which to analyse the last sentence of this paragraph: 'ils vous empoigneront [. . .] le plus tranquillement du monde'.

The adjective 'tranquille' is used repeatedly by the Prologue to convey a subtle hierarchy of meanings in the following extracts (Anouilh, 1946, pp.53, 54, 122) to implicate the helplessness of individuals caught in the coil of tragedy (a2, b2), or the calm unconcern of those who are left to live on after the tragedy (c2).

a.1 That's all it takes. And afterwards, no need to do anything. It does itself. (2000, p.25)
a.2 C'est tout. Après on n'a plus qu'à laisser faire. On est tranquille. Celà roule tout seul.

b.1 But tragedy is so peaceful! For one thing, everybody's on a par. All innocent! (2000, p.26)
b.2 Dans la tragédie, on est tranquille. D'abord on est entre soi. On est tous innocent, en somme.

c.1 So. Antigone was right – it would have been nice and peaceful for us all without her. But now it's over. It's nice and peaceful anyway. (2000, p.60)
c.2 Et voilà. Sans la petite Antigone, c'est vrai, ils auraient tous été bien tranquilles. Mais maintenant, c'est fini. Ils sont tout de même tranquilles.

The translated version omits the phrase 'on est tranquille' in a.1, though it still manages to imply that tragedy absolves the participants from the need to act. In b.2 and c.2, however, the standard implicature of cosy and domestic unconcern generated by 'entre soi' and 'ils sont tout de même tranquilles' does not transfer in the translation. Instead, the translator chooses to emphasize the egalitarianism of participants in the face of tragedy 'on a par' which is an amplification of 'on est tous innocents'.

The use of 'tranquillement' as an adverb also standardly implicates that the guards lack moral stature, going about their daily lives with their petty concerns, like Socrates' contented pigs. They are not the bad guys, but they are not good either, and indeed, they have no moral qualms about manhandling the accused when told so to do: for they are the unthinking agents of oppression. The English translation, by contrast, plays down the ethical dimension, and by turning them into vulgar louts, alienated from Creon's aristocratic Royal Household, seems instead to emphasize class distinctions.

Conventional implicatures are also generated by the use of certain forms such as, for example, the marked use of the present tense to express past action in a narrative context.[10] The messenger's speech is a case in point. By way of reporting the events which have just happened to the Queen, the messenger enables the audience to relive the emotionally charged moments of Antigone's and Haemon's death. Consider the following extract from Anouilh's play (Anouilh, 1946, p.118):

> On venait de jeter Antigone dans son trou. On n'avait pas encore fini de rouler les derniers blocs de pierre lorsque Créon et tous ceux qui l'entourent entendent des plaintes qui sortent soudain du tombeau [. . .] Tous regardent Créon, et lui qui a deviné le premier, lui qui sait déjà avant tous les autres, hurle soudain comme un fou: « Enlevez les pierres ! Enlevez les pierres ! »

I have observed elsewhere (Wolf 2006) that the diegetic present frequently occurs in French native speakers' narratives at points where the event is presented as if there is complete involvement between the speaker and the utterance (see Ducrot's 'engagement énonciatif').[11] In the above example, the use of the diegetic present heightens the sense that Antigone's death is being relived by the messenger with a sense of immediacy rather than just reported to the Queen, and by way of communicative trope, to the audience as well.

Compare with the English translation (2000, p.58):

> They'd just put Antigone in the cave. They hadn't finished rolling the last blocks of stone into place when Creon and all those around him heard cries issuing from the tomb [. . .] They all looked at Creon. And he was the first to guess. He suddenly shrieked like a madman: 'Take away the stones! Take away the stones!'

It is noteworthy that I initially performed the part of the messenger (see note 3), showing immediacy and full involvement with the event being described. This clashed with directorial instructions which were to the effect that the news be reported to the Queen in a detached manner, taking into consideration the momentous and tragic nature of the news whilst showing no direct emotional involvement with it. This is, indeed, how I ended up performing the soliloquy with the attendant consequence that the standardly generated implicature of 'immediacy' was not conveyed to the English speaking audience.

As for the exploitation of maxims, this is illustrated by the frequent floutings of the maxim of quality. Consider the following extract at the beginning of the play when Antigone, with the foreboding of her imminent death, finds child-like comfort in the arms of her nurse who says (Anouilh, 1946, p.32):

La nourrice:	Qu'est-ce que tu as, ma petite colombe?
Antigone:	Rien, nounou. Je suis seulement encore un peu petite pour tout cela. Mais il n'y a que toi qui dois le savoir.
La nourrice:	Trop petite pourquoi, ma mésange?
Antigone:	Pour rien, nounou. Et puis, tu es là. Je tiens ta bonne main rugueuse qui sauve de tout, toujours, je le sais bien. Peut-être qu'elle va me sauver encore. Tu es si puissante, nounou.

The maxim of quality encourages speakers not to say that for which they lack adequate evidence. So in saying that the nurse's hand saves her, Antigone is flouting the maxim of quality, and listeners are likely to derive implicatures along the lines that the nurse is Christ-like, a kind of saviour-nurse, reinforced by the adjective 'puissante' (almighty, all powerful) which has been omitted in the translation (2000, p.15).

Nurse:	What's the matter, my little dove?
Antigone:	Nothing, nan. Just that I'm still a bit too small for it all.
Nurse:	A bit small for what, my sparrow?
Antigone:	Nothing. Anyway, you're here, I'm holding the rough hand that's always kept me safe from everything. Perhaps it will keep me safe still.

In the above translation, Antigone also flouts quality, though the expression 'kept me safe from everything' is unlikely to cause the audience to get in a biblical implicature. The omission of the 'puissante' further deprives the audience of cues to implicature generation which were accessible to the audience of the original and the nurse may consequently be presented as a more secondary character than in Anouilh's version: a motherly English-style nanny.

Another example seems particularly interesting in that it colours our understanding of Creon as the main protagonist of the play. In the original text, Creon contrasts his style of leadership with that of Oedipus, Antigone's father (Anouilh, 1946, p.69):

J'ai résolu, avec moins d'ambition que ton père, de m'employer tout simplement à rendre l'ordre de ce monde un peu moins absurde.

Again, the maxim of quality is flouted: Creon's decision, allegedly less ambitious than that of Oedipus, turns out to be no less than restoring order to a hitherto absurd world. The implicature is achieved that his job, though mundane, is ordained by God. Creon, precisely like Oedipus whom he criticizes, is also a victim of hubris. Consider how much less arrogant, how much more sensible, Creon is portrayed in the translation below (2000, p.33):

> All I aim at now I'm King is *to try* to see the world's a bit more sensibly run.

It may be argued that such a portrayal of Creon as someone ruled by reason and common sense is responsible for critics' comments (Freeman, 2000, p.xli) that he is, 'human and sensitive', 'justified politically ', and 'enhanced by his fortitude in carrying straight on the same day with the thankless task of governing Thebes'. Such observations point to a quasi Austenian contrast between 'sense' and 'sensibility'[12] the former represented by Creon's political authority, and the latter by Antigone's tortured idealism. I believe such views may be based on an insecure grasp of the inferences conveyed by the original text and the related translational phenomena outlined above.

The staging process

As I pointed out earlier, the problematic nature of implicature transfer from one cultural context to another is fairly well documented. A cautionary note is occasionally struck by Malmkjaer (1998, p.37), for example, to the effect that readers, though they may not be able to generate the same sorts of implicatures as those expressed in the original text, might still be able to do so when they take into consideration the co-text of the work. This comment has drawn my attention to how such implicatures might be recovered through the staging process.

Note, for example, how in the following extracts implicatures which were generated in the original text and not recovered in the translation were finally generated through the staging process.

Implicatures standardly generated through conventional lexical meaning included the name 'Floss' for the dog in the English translation (Fr. 'douce'). The incongruity of the name suggestive of sheep rather than a dog was immediately picked on by the actors with the attendant jokes of a sexual nature. Other names were suggested and 'belle' was agreed on which remarkably can be a French dog's name and is very close in terms of standard associations to 'douce' (sweet). Similarly, the word 'girdle' (Fr. 'ceinture'), was replaced by 'belt' on the director's suggestion that the word was obsolete. Directorial instructions also recovered use of the French 'moi, je' to standardly imply contrast, by asking the actor taking the part of Antigone consistently to stress the pronoun 'I' (Anouilh, 1946, p.25):[13]

> Moi, je ne suis pas le roi (But *I* am not the King)
> Moi je ne suis pas obligée (*I* don't have to do what I don't want to)

As far as the flouting of maxims were concerned, the staging process did not seem to have an effect on the recovery of implicatures in the translated text. There was one instance, however, where the lack of transfer led to an ambiguity in the target text which was noticed by the director.

Consider the nurse's question to Antigone at the beginning of the play:

> La nourrice: Tu as un amoureux?
> Antigone: Oui, nourrice, oui, le pauvre. J'ai un amoureux.

Antigone, here, flouts the maxim of quantity, adding more than she needs to be informative. By stating 'j'ai un amoureux', she makes it clear to the audience that she is referring to Haemon, not to the brother who is the real reason for her being up so early.

The reading is more ambiguous in the English text where the maxim of quantity is not flouted, thus causing the director to wonder if Antigone was referring to her brother (2000, p.7):

> Nurse: You mean you've got a sweetheart?
> Antigone: Yes, poor thing.

Finally, in a number of instances, the staging process added inferences which were mostly due to random additions in the translated version. In one excerpt, Creon asked the inarticulate guard to come to the point (2000, p.22):

> Right. Speak . . . What are you afraid of?

The director instructed the actors playing Jonas and Creon to use the pauses in the translated text to imply that Jonas feared for his life. There was no such hesitation in the French version and the focus was not so manifestly on Jonas' fear. Rather, Creon implied that the soldier's inarticulate report was tedious, if not comical, (Anouilh, 1946, p.48):

> C'est bon. Parle. De quoi as tu peur?

Lastly, in the English text, the guards utter snippets like: 'come on, this way!' (see 2000, p.49) when Antigone is taken away. In the original French, however, the guards are unvoiced: they simply obey orders blindly which strengthens the implicature, mentioned earlier, of the guards as blind agents of oppression.

What the examples above tell us about the staging process is that it is, in effect, a re-contextualization of forms which takes inferential meaning into consideration.

This, I believe, takes a similar line to the one advocated by the scholar in theatre studies Patrice Pavis (2000, p.19) in his discussion of *mise en scène*, and it would be worthwhile at this point to offer a brief overview of the main aspects of his approach.

Pavis (2000, p.348) favours a pragmatic approach which allows for words to be recovered in their situational context 'qui parle à qui, dans quel but, avec quels sous-entendus',[14] and *mise en scène* is perceived as a visual recovery of verbal strategies, 'une mise en vue des stratégies de paroles' (Pavis, 2000, p.348).

How is such a *mise en vue* achieved? A disconnection is noted in dramatic discourse between the verbal dimension of the text, 'le dire', and its non verbal staging, 'le faire' (Pavis, 2000, p.19). *Mise en scène*, then, as an embodying, 'une incarnation scénique' (Pavis, 2000, p.19), of 'le dire', enables directors to bridge the gap between the text and its situational context. This process by which directors identify the performative aspects of discourse, Pavis (2000, p.363) refers to as *concrétisation*: 'le metteur en scène concrétise sa lecture de la pièce dans la mise en scène', that is, through the staging process, the director makes his reading of the play explicit and concrete. *Concrétisation*, then, is an interpretative reading of the play made concrete through the bodies of the actors (des corps), and their voices (des voix). It is also a dramaturgical interpretation (une analyse dramaturgique) which, by means of staging, (par la scène), becomes a concrete act of utterance (énonciation).

One example of *concrétisation* provided by Pavis (2000, p.377) is that of the actor's diction. The rhythm, sometimes halted, of the actor's voice is the very link between the enunciation of a particular actor and the text, a link which Pavis describes as the 'mise en voix et en corps du sens textuel', that is the oral and physical staging of textual meaning. The relationship between the text and its staging is generally understood to be a dialectic relationship which obtains between the text and its staging: 'la situation pragmatique de la scène est déterminée par la lecture du texte, laquelle lecture est elle-même influencée par la situation de la scène' (Pavis, 2000, p.26), that is the situational context of the stage is determined by the reading of the text which is in turn influenced by the situational context of staging. And indeed, this is what demonstrably occurred in the staging of *Antigone* referred to here as a re-contextualization of forms.

Furthermore, one important characteristic of texts is their ability to have their meaning infinitely 'reactivated' by what Pavis (2000, p.356) refers to as their social context, 'le texte est concrétisable [. . .] à l'infini'. The discourse related to *mise en scène* is only fully recovered when its products of *concrétisation* have been assimilated by a particular and historically-constrained audience. In this sense, *mise en scènes* are dependent on ideological codes which impose limited interpretations on the text (Pavis, 2000, p.253), so that audiences carry out their own *mise en scènes* 'une mise en scène dans la mise en scène (Pavis, 2000, p.253). The task of criticism is not therefore so much to recover the intentions of the author or the director as to describe the relationship between the text, the staging process and the critical expectations of contemporary audiences confronted by the text.

In describing the relationship between the text, its staging and audience reception, however, it is important to make a distinction, which is not made by Pavis, between what is said by the text and what is implied by it. This is because there are differences in the way meaning is recovered conventionally and conversationally.

Besides, Pavis (2000, p.253), when he talks about clarifying ideological meaning, tends to conflate presuppositions (présupposés), context-independent inferences generated by linguistic meaning only, and 'sous-entendus', inferences generated by the situational context: 'ceci nous oblige à élucider les présupposés idéologiques **contextuels, conventionnels**[15] qui gouvernent nos codes de réception'.

Such a conflation of explicit and implied meaning is problematic mainly because, as we have seen, *mise en scène* finds it more difficult to identify and recover conversational implicatures than conventional inferences, with consequent gains and losses in re-contextualization. This may be because conversational implicatures in translation largely predetermine the dramaturgical and theatrical staging process and, in the words of A. Vitez (1982, p.9), the translation 'contains its [own] *mise en scène*'. Unarguably, the process of translation frequently orients the *mise en scène* in an irreversible direction.[16]

But this is not always the case. As this study has attempted to show, the efficacy of the staging process is such that it can enable the recovery of standardly implicated inferences as well, even when the actors and director have no knowledge of the source text.

It is likely then that an informed *mise en scène*, that is one which takes into consideration the entire co-text of the work and collaboration between writers, directors and translators, would be able to trigger implicature generation at levels beyond that of the standardly implicated.

The idea that translations are irretrievably severed from the original text's intentions is overly pessimistic when we are concerned with performance texts.

Notes

1. Paul Grice, a philosopher of language, is one of the founders of the modern field of pragmatics.
2. An 'implicature' is a type of inference which arises when 'it is possible to mean [. . .] more than what is actually said' (Levinson, 1983, p.97).
3. Methodological note: It must be pointed out that since I performed the part of the 'messenger' in the play under scrutiny, I took advantage of my privileged position as an actor to observe the staging process of the rehearsals by means of field notes. Although I had obtained full consent from the director and the cast to take copious notes of directorial and actors' input, at no stage during the play did I interfere with instructions. What I shall refer to as the 'staging process' or 'mise en scène' includes direction, actors' suggestions and any sort of amendment effected on the performance text.
4. John Lyons (1977) lists criteria of non-canonical utterances which are met by the majority of translated literary texts: 'if they are written rather than spoken [. . .] if the participants in the language event, or the moment of transmission and the moment of reception, are widely separated in space and time; if the participants cannot see one another, or cannot see what the other can see; and so on.'
5. I shall return to this point later in the context of House's (2006) recent article on re-contextualization.
6. The terminology Grice (1975, p.43) introduces as 'terms of art' includes the following: 'to implicate' (cf. to imply), 'implicature' (cf. implying), 'implicatum' (cf. what is implied). Of these, the second is the most commonly used in the literature.
7. For a clear presentation of how conversational explicatures are worked out by speakers and hearers see Levinson (1983, p.113).
8. In the field of linguistic pragmatics, this process of understanding what has been implied is often referred to as 'recovery of meaning' or 'recovery of implicature'.

9. See Catherine Kerbrat-Orecchioni's (1980) analysis of theatrical discourse as fundamentally non-canonical.
10. I shall call this marked use of the present, the diegetic present. The word 'Diegesis' is used here to refer to the domain of 'story time'. In line with Fleischman (1990, p.376), I use the phrase 'diegetic present' (*DPR*) as a cover term for two distinct varieties of PR:

   (i) The 'Historical PR' (*HP*) which is used exclusively in written narratives and occurs in clusters.
   (ii) The 'Narrative PR' (*NP*) which is illustrated by the messenger's speech in Antigone. The *NP* is essentially a phenomenon of orally performed narratives and occurs in alternation with the past.

11. Oswald Ducrot (1984, p.199) distinguishes between the 'locutor as such' (Locutor-L) and the locutor as being in the world (Locutor-λ). An interjection such as 'Pof', implicates the locutor fully, i.e. the locutor-L is seen in her 'engagement énonciatif' (1984, p.200). By contrast, the locutor can express say, sadness, e.g. 'j'étais déprimé'. In this case, it is the locutor-λ who makes a declaration as to his state of unhappiness. The locutor's subjectivity in declarative statements appears to be more external to an utterance than interjections which locate subjectivity right in its midst.
12. *Sense and Sensibility* is the title of the novel, written by Jane Austen, (hence Austenian), which illustrates the contrasting values associated with *sense*, a valuing of reason, and *sensibility*, characterized by a reliance on sentimental behaviour.
13. This was done without any reference to the original text on the part of the director or the actors, just a recovery based on an understanding of the co-text.
14. Pavis draws loosely on the findings of Speech Act theory and Ducrot's (1984) work on théories de l'énonciation.
15. The words 'conventionnels' and 'contextuels' are in bold here to show how Pavis has conflated two notions normally kept very much apart in the field of Pragmatics.
16. Pavis (1992) shows how Porcell's translation of the phrase 'tüchtige Gesellschaft' as 'société efficace' in Strauss' (1986) 'Der Park' led to a French 'mise en scène' of a society concerned with bureaucratic efficiency. In the German staging, by contrast, the focus was on the Germany of the 'petit bourgeois' myth, 'tüchtige' meaning 'hard-working' rather than efficient.

References

Anouilh, J. (1946) *Antigone* (Paris: Editions de la Table Ronde)
Anouilh, J. (2000) *Antigone*, B. Bray (trans.) (London: Methuen Publishing)
Baker, M. (1992) *In Other Words: a Coursebook on Translation* (London: Routledge)
Blakemore, D. (1992) *Understanding Utterances: an Introduction to Pragmatics* (Oxford: Blackwell)

Ducrot, O. (1984) *Le Dire et le Dit* (Paris: Editions de Minuit)
Fleischman, S. (1990) *Tense and Narrativity* (London: Routledge)
Freeman, D. (2000) 'Commentary and notes', in J. Anouilh *Antigone*, B. Bray (trans.) (London: Methuen Publishing)
Gonzalez, L. P. (2006) 'Interpreting strategic recontextualization cues in the courtroom: corpus-based insights into the pragmatic force of non-restrictive relative clauses, *Journal of Pragmatics* (38), pp.390–417
Grice, H. P. (1975) 'Logic and Conversation', in P. Cole and J. L. Morgan (eds.) *Syntax and Semantics 3: Speech Acts* (New York: Academic Press), pp.41–58
Gutt, E. A. (1991) *Translation and Relevance: Cognition and Context* (Oxford: Blackwell)
House, J. (2006) 'Text and context in translation', *Journal of Pragmatics* (38), pp.338–58
Kant, I. (1781) *Immanuel Kant's Critique of Pure Reason*, N. Kemp Smith (trans.) (1933) (London: Macmillan)
Kerbrat-Orecchioni, C. (1980) *L'énonciation de la subjectivité dans le langage* (Paris: Armand Colin)
Levinson, S. C. (1983) *Pragmatics* (Cambridge: Cambridge University Press)
Lyons, J. (1977) *Semantics* (Cambridge: Cambridge University Press)
Malmkjær, K. (1998) 'Cooperation and literary translation', in L. Hickey (ed.) *The Pragmatics of Translation* (Clevedon: Multilingual Matters), pp.25–40
Malmkjær, K. (2005) *Linguistics and the Language of Translation* (Edinburgh: Edinburgh University Press)
Mason, I. (2006) 'On mutual accessibility of contextual assumptions in dialogue interpreting', *Journal of Pragmatics* (38), pp.359–73
Pavis, P. (1992) *Theatre at the Crossroads of Culture*, L. Kruger (trans.) (London: Routledge)
Pavis, P. (2000) *Vers une Théorie de la Pratique Théâtrale, Voix et Images de la Scène 3* (Villeneuve-d'Ascq (nord): Presses Universitaires du Septentrion)
Setton, R. (2006) 'Context in simultaneous interpretation', *Journal of Pragmatics* (38), pp.374–89
Strauss, B. (1986) *Der Park*, C. Porcell (trans.) (Paris: Gallimard)
Toury, G. (1980) 'Translated literature: system, norm, performance: toward a TT-oriented approach to literary translation', in G. Toury (1980) *In Search of a Theory of Translation* (Tel Aviv: The Porter Institute for Poetics and Semiotics, Tel Aviv University), pp.35–50
Vitez, A. (1982) 'Le devoir de traduire', *Theâtre/Public* (44), pp.6–9
Wolf, A. J. E. (2006) *Subjectivity in a Second Language: Conveying the Expression of Self* (Oxford: Peter Lang)

Part II
Practical Perspectives on Translation, Adapting and Staging

6
Translating Bodies: Strategies for Exploiting Embodied Knowledge in the Translation and Adaptation of Chinese *Xiqu* Plays

Megan Evans

In the highly stylized and conventionalized[1] performance forms of Chinese *xiqu* (Chinese opera), the physical and vocal techniques by which the words of a given playscript are conveyed to an audience carry as much, if not more, importance than the words themselves. *Xiqu* performance conventions therefore warrant meaningful representation in the reworking of *xiqu*-derived playscripts for performances in other languages and contexts. Based on his own embodied practice in Asian martial arts and performance forms, Phillip Zarrilli (2004, pp.660–1) urges actor training with 'more positive modes of cultivating the types of bodily awareness often required of the actor/performer'. In her cognitive science based re-evaluation of actor training, Rhonda Blair (2008, pp.25–6) notes the pervasive problem of 'compartmentalization' in acting in the United States by which 'mind is separated from body, feeling from intellect' when actually 'these parts of our selves are inseparable: without the material of the living body, there is no mind, and without feeling there is no true reason'. Even in the verbally dominated modes of performance typically associated with realism, Blair (2008, p.81) concludes that the current state of cognitive research demands new approaches to actor training and rehearsal that link emotional and imagistic processes 'to a detailed kinaesthetic score that supports the body-mapping of . . . images', and that provide sufficient time to get that kinaesthetic score 'into the body' (2008, pp.57–8). Increasingly clear evidence of the efficacy of embodied experience in actor training points to related potential in processes of translation and adaptation. In this article, I examine means of employing direct bodily experience of Chinese *xiqu* performance techniques in processes of 'transposing' *xiqu* plays for English speaking audiences.[2] As a director and performer, my focus is primarily on the development of a *performance text* from a playscript and I am addressing the messy territory at issue in two admittedly simple, but functional ways: I will use 'translation' for English language scripts intended for performance primarily in modes of traditional *xiqu* performance conventions; I will use 'adaptation' for English language scripts intended for other than traditional *xiqu* modes of performance.

There are several ways to attend to the embodied performance text when presenting a *xiqu* play in a live performance for an English speaking audience. One can bring the audience to China, or the Chinese performers to the 'foreign' audience, augmenting the performance text with surtitles or audio commentary in English. This approach resolves many, though certainly not all of the issues raised by translating for a conventionalized form (see Lindsay, 2006, pp.9–16). Alternatively, one can demand that professional *xiqu* performers learn and perform in foreign languages, as occurs frequently in Western opera for the opposite reason of allowing performance of the work in the language for which it was composed and also requiring surtitles for many audiences. This strategy is discussed below in reference to professional *kunqu* actress Qian Yi's performance in an adaptation of the thirteenth-century *xiqu* play *Injustice Done to Dou E*. If one is staging a *xiqu* play using *xiqu* performance conventions but with mostly English speaking performers, as at University of Hawai'i, one must prepare a 'playable' English translation. For purposes of this discussion, a translation that is both amenable to *xiqu* vocal techniques, discussed further below, and able to carry the basic dramatic action can be considered 'playable'. In contrast, if one decides to stage an adaptation of the play in another mode of performance, there are important benefits to be gained from using traditional *xiqu* performance texts as a rehearsal tool, as I did in staging an adaptation of *Injustice Done to Dou E* at Victoria University of Wellington, New Zealand. After a brief introduction to *xiqu*, including a discussion of key theoretical aspects of *xiqu* performance-audience relationships, I discuss several of these strategies in detail.

Chinese *xiqu*

A key aesthetic goal of Chinese *xiqu* is the effective synthesis of numerous performance elements including song, speech, dance-acting, combat, acrobatics, and complex musical structures. Character and plot serve to contextualize emotional expression that the performer enacts through exhibition of highly stylized performance skills achieved through years of training. Moments of heightened emotion are expanded in time and space through vocal and physical stylizations. The *xi* of *xiqu* means play or drama while *qu* means song or music, hence a recent trend to translate *xiqu* as Chinese 'music-drama'. But because of the aesthetic emphasis on synthesis, *xiqu* is a total theatre form that is dependent on embodied expression through *physical* as well as vocal technique.[3] The written scripts of *xiqu* plays represent a much smaller percentage of the performance text than 'spoken drama' (*huaju*). A *xiqu* play requiring two to three hours of playing time might average a half or even a quarter as many written characters as a 'spoken drama' (*huaju*) playscript requiring equivalent playing time (Wichmann, 1991, p.24).

Currently there exist over 300 different regional forms of *xiqu*, distinguished primarily by local musical structures and dialects. Three *xiqu* forms are discussed below. *Jingju* (Beijing/Peking opera) developed in the late eighteenth century and achieved international recognition through the 1930s tours of master performer Mei Lanfang. *Jingju's* Ming Dynasty (1368–1644) predecessor *kunqu* (Kun opera) is enjoying a current resurgence in the wake of its 2001 UNESCO designation as a 'Masterpiece of the Oral and Intangible Heritage of Humanity'. Yuan *zaju* ('variety drama') matured in the Yuan Dynasty (1271–1368) and has many extant playscripts but little remaining evidence of its staging techniques. Within China, a given *xiqu* performance is first and foremost an example of that particular *xiqu* form[4] and secondarily a performance of a particular play within that form. As Wichmann (1991, p.105) notes: 'The source of the example, and therefore of the form itself, is the well-trained actor. Artistry – both for the continued existence of the form and for creation in it – resides in the body, mind, and psyche of the experienced actor.' *Xiqu* plays are frequently adapted to suit specific strengths of a particular group of performers. Though contemporary practice is changing with the increasing importance of directors, playwrights, composers and choreographers in the creative process, *xiqu* performers traditionally have been principally responsible for important aspects of musical composition and staging. Thus a *xiqu* playscript exists primarily to serve the performance conventions of the form. In translating a *xiqu* play for staging in traditional modes of performance, the translator must address those conventions. If, in contrast, the project involves adaptation for a different mode of performance, important artistic and pedagogical advantages can still be gained by including embodied experience of *xiqu* performance conventions in the project.

Actor–character–audience relationships in *xiqu*

Central to this discussion is the conceptual relationship between the physical bodies of the performers on a *xiqu* stage, the fictional world being portrayed, and the audience reception of these. Because of its influence on Bertolt Brecht, *xiqu* sits at a crucial intersection of Asian and Anglo-European theorization of these relationships. In his 1935 viewing of a performance by master *xiqu* artist Mei Lanfang, Brecht perceived an aesthetic distance between performer and character that disrupted empathetic engagement by the spectator, prompting his first published articulation of the *Verfremdungseffekt* in the essay 'On Chinese Acting' (Martin, 2000, p.5). In the 1980s, noted *xiqu* artist and scholar A Jia theorized *xiqu* acting in reference to Brecht as a process by which both actor and spectator empathize with the character to some extent while simultaneously maintaining the intellectual distance necessary to perform or appreciate the performance skills employed for a particular role. A Jia argued (in Liu Yizhen 1988, pp.127–8), 'when an audience watches a play, there is both intellectual judgment and emotional response – and there is also appreciation of beauty. This is the method of communication between *xiqu* and the audience.' In the 1990s, Min Tian (1997, p.206) questioned the coherence A Jia found between *xiqu* and Brechtian modes of aesthetic communication, arguing persuasively that *xiqu* aesthetics depend primarily on audience *familiarity* with *xiqu* conventions and performance skills in a manner fundamentally different than Brecht's goal of making the familiar seem strange. However, foreign audiences (and many contemporary Chinese viewers) will be unfamiliar with *xiqu* conventions and will be likely to have an experience closer to Brecht's than that theorized by Tian for a 'traditional' *xiqu* audience.

More recently, Haiping Yan (2003, p.75), drawing on theoretical writings of playwright and Nobel prize winner Gao Xingjian, has argued for the label 'suppositionality' to describe a confluence of aesthetic and ethical dialectics in *xiqu* acting and dramaturgy that operates quite differently than Brecht's understanding of distancing in *xiqu* yet still activates 'transformative imaginations' that engage the spectator's moral and ethical reading of the staged events in relation to his or her own experience. Yan notes that the 'observable effect of theatricality' in a *xiqu* performance is a process by which 'what there is to know' is only pointed at, such that an actor carrying a stylized horsewhip prop and raising his leg 'points to' the existence of a horse the character is riding. Yan argues (2003, p.84) that this 'pointing at what there is to know' necessarily presumes an 'active participatory role for audiences, for they engage in the pointing by doing their own knowing and feeling'. On the other hand, she argues that *xiqu's* external theatricality cannot be separated, as Brecht did, from the internal cultural dynamics through which a *xiqu* spectator casts a portrayal of injustice against their own understanding of what *should be* and decides how to feel about what is being 'pointed at' onstage. Yan (2003, p.86) argues that:

> these are not the usual kind of subconscious identifications that mark and mask the absence of agency à la Western drama. The 'moving and moved feelings' that realize the aesthetic ideal of Chinese music-dramas are energies knowingly gestured, gathered, felt, imagined, and inhabited by both performers and audiences. They are living sites of theatricality in Chinese performing art.

Certainly, familiarity with *xiqu* performance conventions will give a viewer more complete access to potential meanings generated by a *xiqu* performance. From my years of experience introducing students to *xiqu*, however, I know that many conventions are accessible to foreign spectators (an oar and rowing action to indicate boat travel), while others can quickly be made accessible (once viewers learn to recognize the horsewhip prop, the conventional movements of horse travel become legible). Yan's formulation does not theoretically preclude a foreign audience member from achieving some level of suppositional contribution to these 'moving and moved feelings'. Indeed, the degree to which such an aesthetic ideal is achieved can serve as an important indicator of the 'success' of the staging of a particular translation for an individual spectator.

As described above by A Jia, and from my own experience as frequent spectator of professional *xiqu* performance, effective operation of this knowing confluence of feelings and judgments demands the *skilful* execution of *xiqu* performance techniques. And when these skills are deployed on the stage, the performer's body becomes an activated, positive force even if, and often precisely because, the character is in a circumstance of extreme powerlessness. In such circumstances, the exhibition of *xiqu* skill arguably stages reflections of 'what should be': a powerless victim of injustice expresses her rage through artful deployment of vocal and physical prowess. The embodied potency of extraordinary skill exhibited in a successful *xiqu* performance itself expands the boundaries of 'what there is to know' both in terms of how the actor feels about the character's situation and how the audience responds to the performance.

Yan's formulation (2003, p.86) of 'moving and moved feelings ... *knowingly* inhabited by both performers and audiences' is supported by developments in cognitive science and phenomenological research, which make increasingly clear that bodies have ways of knowing fundamentally different from those of brains. Infants learn how to be in the world first by learning how to move. As leading dance theorist and cognitive philosopher Maxine Sheets-Johnstone (1999, p.57) notes, the qualitative characters of our own movements:

> are not a 'mental product,' but the product of animation. They are created by the movement itself ... When [as infants] we learn to move ourselves, we learn to distinguish just such kinetic bodily feelings as smoothness and clumsiness, swiftness and slowness, brusqueness and gentleness, not in so many words, but in so many bodily-felt distinctions.

This kinesthetic sense is also empathetic – learning how to move ourselves allows us to 'feel' what another is moving (Parviainen, 2002, pp.20–1). Similarly with regard to vocal performance, cognitive research supports the conclusion that self-awareness is sustained by the embodied language of speech, and activates fundamentally different processes than language that is only thought or heard (Stamenov, 2002, pp.30–2). These developments suggest that embodied experience of *xiqu* performance conventions can be exploited in the process of translation or adaptation to achieve results fundamentally different and arguably more stageworthy than those produced through attention to the words of a *xiqu* playscript alone. In the following discussion, I examine strategies for exploiting such embodied experiences in the rehearsal process of translated and adapted *xiqu* plays.

An embodied approach to translation: University of Hawai'i Asian theatre program

The University of Hawai'i at Mānoa (UH) Asian Theatre Program has had an ongoing and extraordinarily fruitful collaboration with the Jiangsu Province *Jingju* Company that has facilitated rigorous embodied training of hundreds of UH students in *jingju* performance techniques and staged full productions of five English language *jingju* plays[5] for thousands of spectators. These projects have been carried out under the overall artistic direction of master *jingju* performer and teacher Madame

Shen Xiaomei, working closely with UH Professor Elizabeth Wichmann-Walczak. Madame Shen is Mei Lanfang's youngest living disciple and has taught the female roles on several of the projects.[6] I participated in three of these residencies as a student performer, and in a fourth project as an interpreter for group and individual classes.

In the first phase of the project and while still in China, Madam Shen and the other Chinese teachers develop a version of the performance text of the chosen play, taking into account such things as playing time, technical difficulty of sung passages and movement sequences, and clarity of the story for Hawai'i audiences. To prepare the English translation, Wichmann-Walczak works primarily from audio recordings of the teachers performing the stylized speech and arias they have set. Relying on her own extensive training in *jingju* performance conventions, she strives to maintain the syllabic rhythms of sections of stylized speech and rigorously follows the phrasing, syllable count, and internal rhyme schemes of the original Chinese aria lyrics with the goal of producing English phrases that are 'playable' using *jingju* vocal conventions (Wichmann, 1994, pp.101–2).[7] In addition, *jingju* vocal technique demands precise vocal placement and resonance that differs between the various role types and serves as an important marker of a character's gender, age, and even key psychological traits. There is a common saying among *jingju* artists that performers of young female roles (who utilize a bright, forwardly-placed head resonation called 'falsetto'[8] or 'small voice') love the sound [i:] as in 'see' and fear the sound [a:] as in 'father', while for performers of older male and female roles (who utilize a vocal placement mixing soft palate and chest resonation called 'real' or 'large voice') the reverse is true. Scripts prepared by authors who understand *jingju* singing conventions will take this into account and attempt to phrase lyrics so that ornamented singing will land on the appropriate or related vowel. In preparing the English translations for the UH *jingju* productions, Wichmann-Walczak also tries to match as closely as possible the particular vowel sound of the original whenever a single syllable is sung over a number of notes or held for an extended time. For example, 'liu gai wen yu wei li qing' is a lyric for a young female role in the play *Judge Bao and the Case of Qin Xianglian* (Shen 2001), discussed further below. The line breaks musically and in meaning after the fourth syllable, 'yu', and places the most ornamented singing on the favoured [i:] sound of the syllable 'li'. Wichmann-Walczak's translation – 'I thought you would dry my tears' – also breaks sensically after the fourth syllable and places the ornamented singing on the [i:]-related vowel of 'my'.

From the standpoint of singing, a crucial difference between Mandarin and English is that all final sounds of Mandarin words are voiced – either vowels or the singable consonants [n] as in 'ma<u>n</u>' or [ŋ] as in 'si<u>ng</u>'. In contrast, of course, many English words end on one or more plosive consonants. Wichmann-Walczak avoids these sounds where possible and has actors soften remaining abrasive endings, shifting reliance for audience comprehension to the projected side titles used for all sung passages. Side titles are also used for *xiqu* performances in China to assist comprehension between dialects, and over highly ornamented sung phrases.[9]

For the UH student performers, the first phase of each project begins with physical training. In the spring semester preceding the arrival of the Chinese teachers a course is offered in basic movement techniques for *jingju* involving martial arts kicks, basic stage steps and postures, and set patterns of gestures. A fundamental *xiqu* aesthetic demands that all aspects of vocal and physical performance skill appear beautiful, which in *xiqu* terms means that they appear round and effortless. The basic physical training helps performers control breath, and begin to develop the sensorimotor awareness necessary for *xiqu* performance, in particular a still and lifted upper body that gives the appearance of effortlessness.

The Chinese teachers arrive in the fall for a seven month residency, conducting group movement and voice classes and individual teaching sessions in which performers learn their specific roles. One of the most important strategies Wichmann-Walczak adopts for incorporating embodied experience of *jingju* performance conventions is her practice of providing two or more alternate translations for aria lyrics and passages of heightened poetic recitation. *Jingju* vocal and/or physical stylizations are particularly heightened in these sections, thus presenting the most potential difficulties to achieving a 'playable' English translation. For the same reason, student actors first study these sections of the play in Chinese so that they are attempting to imitate their teachers as exactly as possible. Once basic melodies and sections of recitation are learned, the student performers begin to make the shift to English and to learn the accompanying movement. Though an interpreter is present in the individual teaching sessions, the student performers have primary responsibility for developing the most effective combination between the English language options and the performance text they have learned from their teachers.

Figure 6.1 is a copy of a page from my rehearsal script for *Judge Bao and the Case of Qin Xianglian* (Shen Xiaomei et al., 2002). I played the role of Qin Xianglian, abandoned by her husband, a poor scholar who has since placed first in the Imperial examination and married

Figure 6.1: Author's rehearsal script of Elizabeth Wichmann-Walczak's working translation of the traditional *jingju* play *Judge Bao and the Case of Qin Xianglian* (Shen Xiaomei et al., 2002), performed at Kennedy Theatre, University of Hawai'i at Mānoa, February 2002

the Emperor's sister by concealing his first marriage. Qin Xianglian gains access to his palace and begs him to acknowledge her and their children. He denies her, knowing he would be executed for lying to the imperial family.

As evident from Figure 6.1, Wichmann-Walczak's working translations facilitate performer experimentation with English language options and *jingju* vocal stylizations in the following ways:

1. Graphic spacing of English syllables above the original Chinese aids rhythmic and melodic precision;
2. Breaks in English syllables are set so as to minimize plosive consonant endings;
3. Options are offered in sets to maintain 'rhyme scheme' such that if the performers choose option (a) ending with 'limb' in the first line of recitation in Figure 6.1, option (a), ending with 'phoenix', in the second line must also be used.

In *jingju,* all vowel sounds are divided into thirteen sets of related sounds. If central vowels of two words are from the same set, they are deemed to rhyme, regardless of final consonants. Wichmann-Walczak similarly works to match only the internal vowel to establish 'rhyming' sets.

In rehearsing the first two lines of recitation, I had no preference between option (a) 'limb' or option (b) 'home' since neither option contained the falsetto-favored [i:] nor feared [a:] and both ended on the easily stylized [m] sound. Tim Wilder playing Chen Shimei using 'large voice' placement, however, found the feared [i:] as well as the final plosive [ks] sound in phoenix particularly difficult for the ornate stylization set for the final word of his line. Recognizing that option (a) ('The finest sparrow suits not a phoenix') more clearly communicated the status distinction at the core of the original metaphor, Wilder settled on 'How could *this* phoenix nest with a swallow.' For the second exchange Wilder and I agreed that the directness of option (c) ('How is love forgot? / Of the past speak not') served the building tempo well. For the third exchange, my teacher, Li Zhenghua, had set a forceful counter-cross combined with an elaborately stylized final syllable in the strongest accusation Qin makes to Chen in the entire scene. The movement and vocal stylization fit easily with the falsetto-favored [i:] of the Chinese word 'qi', but I struggled to make it work on the English options 'truth' and 'self'. After discussion with Wichmann-Walczak, I finally revised the phrase to a version I found more playable: 'deceiving others, one's self is deceived'. Because Wilder's vocal stylization for the final line did not include an extended syllable, he agreed to the ending [i:] sound of 'Wealth bests poverty, you've no chance with me.'

Though many of the UH student performers do not speak Chinese and have not previously encountered *jingju*, through this process they explore aural qualities of their performance text in Chinese. Since they carry primary responsibility for negotiating the relationship between two or more options in English phrasing with the vocal stylizations and specific gestures set by the teachers, they take important authority in the final revisions of a playscript tailored to their individual, if imperfect, *jingju* performance skills.

Embodied approaches to *xiqu* adaptation: two productions of Charles Mee's playscript *Utopian Highway* inspired by *Injustice Done to Dou E* by Guan Hanqing (c. 1241–1320)

Where a project initiates from a *xiqu* playscript but *xiqu* performance conventions will not be the primary mode of staging, artistic and pedagogical impulses may still support inclusion of meaningful attention to these conventions in some way. Cognitive research suggests that to be most meaningful, such attention should include embodied experience. For reasons discussed below, embodied experience of *xiqu* performance conventions is a particularly appropriate preparation for staging an adaptation of the Yuan Dynasty (1271–1368) *zaju* play *Injustice Done to Dou E (Dou E yuan;* Dou E is the heroine's name and is pronounced [tǝu] as in dough, ['ǝ] similar to 'uh'), by Guan Hanqing (c. 1241–1320), also known as *Snow in Midsummer,* and *Snow in June.*

Guan Hanqing's script is permeated with images of physical violence calling attention to the physical bodies of the performers engaged in its staging. Every major plot point involves a threat or occurrence of bodily harm. After a brief prologue, Dou E's adopted mother, Mistress Cai, is nearly strangled by a debtor trying to escape his debt. Later, Dou E is falsely accused of murder and brought to trial before a corrupt official. She professes her innocence then sings lyrics describing the torture she endures (Guan date uncertain, trans. Qian Ma, 2005, p.68):

> No sooner does the beating pause and hardly do I revive,
> Than once again I fall unconscious.
> One thousand beatings and ten thousand tortures,
> One blow down, one streak of blood, and one strip of skin.
> I am beaten till pieces of flesh fly in every direction
> And I am dripping with blood,
> Who knows the wrongs I feel in my heart?

Dou E only confesses when the corrupt official threatens to have her adoptive mother beaten instead. At her execution, Dou E calls on Heaven to prove her innocence by bringing heavy snowfall to cover her corpse, even though it is hot summer.

Images of fragile, mortal bodies damaged by lies and corruption permeate this play perhaps in response to the brutality of the Mongol invasion, and ensuing enslavement and injustice suffered by conquered Han Chinese,[10] though such brutal bodily references make frequent appearances across the *xiqu* repertoire. Language describing physical brutality calls attention to the physical bodies of performers speaking or singing that language and to the performance skills through which that language is embodied onstage. While little evidence remains about specific performance techniques used to stage Yuan *zaju* plays, existing *xiqu* forms share at least the following fundamental similarities with *zaju*: an established structure of role types as the framework around which the plot is organized; the importance of song, lyrics for which often include the most complete expression of the emotions of central characters; and, the use of stylized movement patterns (Dolby, 1988 pp.50–3; Crump, 1980, pp.72–3). These similarities support incorporation of extant *xiqu* performance techniques in non-*xiqu* adaptations using professional *xiqu* performers. They also suggest that embodied study of such techniques in rehearsal can be a useful way to deepen student performer understanding of Guan's thirteenth-century playscript while also facilitating new embodied perceptions of the roles they must develop to stage a non-*xiqu* adaptation.

US playwright Charles Mee was commissioned by director Chen Shizheng to adapt *Injustice Done to Dou E* for a production at the American Repertory Theatre (ART) in 2003. Mee, who describes this work as 'inspired by' rather than 'an adaptation of' an early folk tale and Guan's Yuan era dramatization, set his version in a New York-esque urban America and titled it *Utopia Parkway*. He carries forward some of the brutality of Guan's original in a chorus ('Street talk part 2') in which a 'bum' is lit on fire 'and left burning through the night' and in the Trial Scene (scene 10) in which various torture techniques are described, including being 'held down in the water, fed human flesh, arms and legs broken, eyes burned out with cigarettes' (Mee 2003). Director Chen reverted to the alternate Chinese title *Snow in June*, but relocated the plot to a stylized rural American south. He commissioned composer Paul Dresher whose score drew heavily on US folk traditions except for the final scene, which used traditional *kunqu* arias, sung in Chinese by the professional *kunqu* performer playing the leading role. While the cultural dislocations operating in this production are unquestionably extreme, they mirror the centuries of borrowing and adaptation of plays from one *xiqu* form to another, with basic plot elements and characters receiving extensive alteration. For example, some twentieth-century *jingju* versions of Guan's play omit the supernatural element altogether: Dou E's father arrives in time to prevent her execution.

In Mee's text, Dou E's father is eliminated from the play. Dou E still calls for snow in summer as proof of her innocence. After her execution, however, her ghost rises to take revenge directly, killing her accuser, the judge and even her adopted mother whose previous actions contributed to the injustice. Guan Hanqing's Dou E does not kill, but she articulates her rage against profound injustice with such force that Heaven and Earth are moved to act on her behalf. In the more secular, diffusely governed world in which Mee's heroine operates, snow still falls to prove her innocence but she must take revenge with her own hands.

First embodied approach: staging developed with a professional *xiqu* performer

Chen Shizheng's staging at ART inscribed this action with an important shift of performance mode. Dou E was played by professional *kunqu* actress Qian Yi, who also starred in the 20-hour 1999 Lincoln Centre *Peony Pavilion* which Chen directed. Working in newly acquired modes of Western experimental music-theatre through most of the production, Qian Yi in a sense embodied the cultural dislocations of the project. A *New York Times* reviewer thought she played with 'haunting, electrifying exoticism' and called her speaking of the English text 'deliciously odd' (Weber 2003). The *TheatreMania* reviewer found it 'odd to hear her delicate *kunju*[11] operatic technique applied to Dresher's homespun ditties; she appears to be singing phonetically and there's a trace of Shirley Temple in her faux-naïve posing' (MacDonald 2003). In the final scene, when the ghost girl rises to take revenge after her execution, Qian Yi shifted to the performance modes of traditional *kunqu*, using an extended aria lifted directly from a *kunqu* version of *Injustice Done to Dou E* that Qian Yi studied during her pre-professional training (Clay 2003; Shanghai Kunju Troupe n.d.). Employing Qian Yi's exceptional *kunqu* performance skills and native language, supported by projected English surtitles, the stylistic shift endowed the ghost with a previously untapped performative strength, which one reviewer found 'particularly wonderful' (MacDonald 2003). At the same time, it staged a removal of the action to *kunqu's* cultural source. For North American spectators unfamiliar with *kunqu* performance conventions this perhaps caused a simultaneous rupture in identifications with the character that had previously been portrayed in accessible US folk melodies. Stylistic discordances caused the *New York Times* reviewer to marvel overall at the production's 'eccentricity' but question 'whether Mr. Chen's culture clash yields something more meaningful' (Weber 2003). Yet this final, and most extreme, 'relocation' clearly served the dramaturgical shift to a poetic borderland between life and death, and staged a kind of revenge of *xiqu* tradition over avant-garde eclecticism.

Second embodied approach: replication as rehearsal tool for student performers

A professional *xiqu* skill level is clearly not attainable in the course of a single rehearsal period, or even over the extended residencies of the UH model. Nonetheless, for student performers seeking developed awareness of performance processes, valuable embodied experience can be achieved through work on brief replication scenes in rehearsal. I recently directed a production using Mee's text *Utopian Parkway* and Dresher's score at Victoria University of Wellington, New Zealand with student actors who had no previous exposure to *xiqu*. (Borrowing a line from Mee's text, we chose the title *Snow in Sweet Summer* since New Zealand has no parkways and commonly sees snow in its winter month of June.) Knowing none of these students could sing the final *kunqu* aria at a performance-worthy level, I had a music student, Tian'er Yao, compose a new closing aria in Western classical music style. For similar reasons, we used *xiqu*-specific gestures and vocal patterns only minimally. On the other hand, we used underlying *xiqu* principles extensively in the production, including: integration of music with onstage action, precisely scored rhythmic transitions, and clarity of emotional progression supported by precise physical score. As part of the rehearsal process, students explored the martial arts kicks and gesture patterns of *jingju* basic training and studied the scene from *Qin Xianglian* shown in Figure 6.1. I used this scene because I have performed it myself, used it effectively in previous teaching, and it had thematic links since Xianglian, like Dou E, also suffered grave injustice.[12] As staged for the UH production, this compact two-character scene integrates major elements of *xiqu* performance. Each role has a brief aria, stylized speech combined with movement, and intense emotional content. Despite the minimal available time, study of the scene gave students embodied appreciation of how *xiqu* performers integrate conventional performance modes with character expression, how vocal and physical stylizations are integrated, how percussion supports and expands a gesture or an expression of emotion. Building on this embodied experience with traditional *xiqu*, we returned continually to these structural principles in developing our staging of *Snow in Sweet Summer*. Following the public performance of *Snow* we revived the replication scene in our final class meeting. Despite not having worked the scene for over two weeks (with all the intervening stress of production), I observed a marked increase in students' confidence and precision and in their ability to imbue *xiqu* gestures and vocal stylizations with emotional force. I believe this observable increase in skill level is evidence of the interchange of embodied knowledges achieved through this rehearsal strategy. Originating out of their tentative performance of *jingju* basic training exercises and the replication scene, the students' nascent understanding of *jingju's* general principles and performance conventions was honed and solidified through the crucible of public performance. This maturing embodied understanding was then applied in their subsequent re-encounter with the performance text of the replication scene to positive effect.

Conclusion

Each of the strategies discussed above makes the embodied experience of *xiqu* performance conventions meaningfully present in the project of translation or adaptation. As Yan Haiping argues, in its traditional setting *xiqu's* suppositional actor–audience relationship requires 'moving and moved feelings . . . knowingly gathered' in an interchange of energies between skilled performers and knowledgeable spectators based on mutual understanding of *xiqu* performance conventions. In each of the cases discussed above such knowledge was presumably not held by a majority of spectators and, indeed is not held by many potential spectators in contemporary China. Only in the ART production did the performer, Qian Yi, possess a professional level of 'fluency' in these conventions, the power of which, when unleashed at the end of the production as she sang traditional arias in Chinese, perhaps upstaged the newly concocted performance text that preceded it. Nonetheless, as judged by the production's critical reception, Mee's adapted playscript coupled with Chen Shizheng's eclectic performance text triggered the confluence of empathetic, aesthetic and ethical responses supportive of 'successful' audience contribution to a *xiqu* performance. In circumstances where performers are not familiar with *xiqu* conventions, whether in a professional or educational setting, the project of developing a performance text for a playscript derived from one of the acknowledged masterpieces of Chinese dramatic literature, such as *Injustice Done to Dou E*, presents an important opportunity to expand awareness of the societal forces and artistic traditions producing that playscript. To sidestep such an opportunity in this age of globalizing cultural flows seems to me foolish, if not irresponsible. Reading, discussion and video viewing offer ready intellectual access to traditional *xiqu* performance conventions. Embodied experience offers a qualitatively different method that is more directly applicable to the project of developing a performance text.

As shown by the UH model, even extensive familiarity with the specific medium of *jingju* performance techniques, such as that exhibited by translator Wichmann-Walczak, can be enhanced by exploiting student performers' embodied knowledge of the performance text obtained through rigorous interaction with their master teachers. Similarly, actors in a production of an adapted *xiqu* play can only benefit from a basic embodied knowledge of the performance conventions for which a source playscript was created. Both the UH model and the staging we developed for *Snow in Sweet Summer* followed an approach similar to that urged by Blair (2008, pp.57–8) in which a precise physical score was developed over a period extensive enough to get it 'into' the performers' bodies. Indeed, consistency and precision is the necessary by-product of incorporating percussion as aural punctuation for speech and movement that is a core aesthetic principle of traditional *xiqu* used in the UH productions and which I adapted to *Snow in Sweet Summer*. From my own training and teaching experience, I know that *jingju* basic physical training can be effectively combined with embodied study of a brief scene and that it is worth the effort. The process expands 'what there is [for a performer] to know' and grounds experimentation toward a staging of an adaptation in an embodied understanding of *xiqu's* artistic potency.

Notes

1. 'Conventionalization' is a direct translation of the term *(chengshihua)* used by *xiqu* artists and scholars to describe the concept of intergenerational transmission of performance rules and patterns (Li and Jiang, 2000, p.69). I prefer it to Barba's universalizing term 'codification' both for its cultural specificity and because the '-ize' suffix invokes attention to ongoing processes, allowing more ready appreciation of innovation in this living tradition.
2. Though the case studies analysed here involve *xiqu* 'source' plays transposed for English-speaking actors and audiences, many of the embodied approaches discussed here would be as equally applicable to 'transpositions' of other conventionalized forms and other language groups.

3. Distinguished from this indigenous music-drama, western-style realistic drama, called *huaju* or spoken drama, did not arrive in China until the early twentieth century.
4. When productions are developed for international touring this formal 'representative' status can become even more critical, as happened with the 1997 Shanghai Ministry of Culture cancellation of *The Peony Pavilion* directed by Chen Shizheng and bound for Lincoln Centre in New York. The production was ultimately staged at the 1998 Lincoln Centre festival and toured successfully in Europe (Jain, 2002, p.122).
5. The sixth is slated for 2010.
6. Other Jiangsu Province Jingju Company members who have participated in these projects are: Mr Shen Fuqing, Mr Lu Genzhang, Ms Li Zhenghua, Ms Zhang Ling and Mr Zhang Xigui.
7. It is worth noting that by ignoring these principles or the many additional requirements for tonal patterns not applicable to English, it is entirely possible for a playwright to produce a *Chinese* language script that is not 'playable' in *jingju* vocal conventions.
8. The Mandarin 'jia sangzi' means 'fake voice'. Since there is no conceptualized difference between how male and female *jingju* performers produce the sound, the concept is somewhat different than in western usage where the term is usually limited to male performers.
9. Since contemporary Chinese audiences are less familiar with the many archaic pronunciations demanded by *jingju* performance conventions, sidetitles are increasingly used for spoken as well as sung sections. While aiding comprehension, the practice creates extended periods of split focus distracting from visual aspects of the performance text and have not been used for spoken sections in the UH production which utilize vernacular English.
10. For statistics about slavery, slaughter and judicial corruption in the Yuan period, see Liu, 1972, p.9.
11. *Kunju* is an alternate name for *kunqu*.
12. Video of professional performances can supplement or even substitute for 'live' teaching if necessary since the goal is pedagogical rather than achieving a performance-worthy level of skill. The in-depth analysis required to 'lift' choreography and vocal stylization from a video recording provokes much deeper learning for student performers than passive video viewing alone. The practice parallels contemporary training in China since audio and video recording devices are increasingly employed by professional *xiqu* performers-in-training to facilitate independent review of material between meetings with their teachers.

References

Blair, R. (2008) *The Actor, Image, and Action: Acting and Cognitive Neuroscience* (London and New York: Routledge)

Clay, C. (December 2003) 'Winter wonderland *Snow in June* is a multicultural treat', *Boston Phoenix* (online) Available at: http://www.bostonphoenix.com/boston/arts/theater/documents/03396397.asp [Accessed 23 October, 2009], pp.12–18

Crump, J. I. (1980) *Chinese Theater in the Days of Kublai Khan* (Tuscon, Arizona: University of Arizona Press)

Dolby, W. (1988) 'Yuan Drama', in C. Mackerras (ed.) *Chinese Theater From Its Origin to the Present Day* (Honolulu: University of Hawai'i Press), pp.32–59

Guan Hanqing (2005) *Dou E Yuan (Injustice Done to Dou E* also known as *Snow in June* and *Snow in Midsummer)* in Ma, Qian (trans.) *Women in Traditional Chinese Theatre: the Heroine's Play* (Lanham: University Press of America, Inc.)

Jain, S. Pertel (2002) 'Contemplating Peonies: A Symposium on Three Productions of Tang Xianzu's *Peony Pavilion'*, *Asian Theatre Journal* (19: 1), pp.121–3

Li, Ruru and Jiang, D. W. (2000) 'Conventionalization: the Soul of Jingju', in R. Mock (ed.) *Performing Processes* (Bristol and Oregon: Intellect), pp.69–82

Lindsay, J. (2006) 'Introduction' in J. Lindsay (ed.,) *Between Tongues: Translation and/of/in Performance in Asia* (Singapore: Singapore University Press), pp.xi–xv

Liu Jung-en (1972) 'Introduction', in Liu Jung-en (trans.) *Six Yuan Plays* (London: Penguin Books Inc.), pp.7–35.

Liu Yizhen (1988) 'Ah Jia's Theory of Xiqu Performance', Hu Dongsheng et al. (trans.), *Asian Theatre Journal* (5: 2), pp.111–31

Ma, Qian (2005) *Women in Traditional Chinese Theatre: The Heroine's Play* (Lanham, MD: University Press of America)

MacDonald, Sandy (2003) 'Theatre Review: *Snow in June'*, *Theatre Mania* 8 December (online), Available at http://www.theatermania.com/content/news.cfm?int_news_id=4173 [Accessed 31 October, 2009]

Martin, C. and H. Bial (2000) 'Introduction', in C. Martin and H. Bial (eds.) *Brecht Sourcebook* (London and New York: Routledge), pp.1–14.

Mee, C. (2003) *Utopia Parkway* (online) Available at: http://www.charlesmee.org/html/utopia.html [Accessed 31 October, 2009]

Mee, C. (2007). Unpublished email interview.

Parviainen, J. (2002) 'Bodily Knowledge: Epistemological Reflections on Dance', *Dance Research Journal* (34: 1), pp.11–26

Shanghai Kunju Troupe (n.d.) 'The Execution of Dou E' on DVD volume with other excerpts from *China's Kunqu Opera Audio-Visual Collection* IRSC CN-E04-04-0010-0/V.J8 (Shanghai: Shanghai Audio-Visual Publishers)

Sheets-Johnstone, M. (1999) *The Primacy of Movement* (Amsterdam: John Benjamins)

Shen Xiaomei, et al. (2002) (unpublished) *Judge Bao and the Case of Qin Xianglian (Qin Xianglian)*: A traditional *jingju* play as scripted, arranged and taught by Madam Shen Xiaomei, Mr Shen Fuqing, Mr Lu Genzhang, and Ms Li Zhenghua. Translated for performance by Elizabeth Wichmann-Walczak and Hui-Mei Chang

Stamenov, M. I. (2002) 'Body schema, body image, and mirror neurons', in H. De Preester, and V. Knockaert (eds.) *Body Image and Body Schema* (Amsterdam: John Benjamins Publishing Company), pp.21–44

Tian, Min (1997) '"Alienation-Effect" for Whom? Brecht's (Mis)interpretation of the Classical Chinese Theatre', *Asian Theatre* (14: 2), pp.200–22

Weber, Bruce (20 December, 2003) 'Theatre Review: A Clash of the Seasons Inspires a Meeting of Cultures', *New York Times* (online) Available at: http://theater2.nytimes.com/mem/theater/treview.html?_r=1&res=9407E6DF153FF933A15751C1A9659C8B63 [Accessed 31 October, 2009]

Wichmann, E. (1991) *Listening to Theatre: the Aural Dimension of Beijing Opera* (Honolulu: University of Hawai'i Press)

Wichmann, E. (1994) '*Xiqu* research and Translation with the Artists in Mind,' *Asian Theatre Journal* (11: 1), pp.97–103
Yan, Haiping (2003) 'Theatricality in Classical Chinese drama', in T. Davis and T. Postlewait (eds.) *Theatricality* (Cambridge: Cambridge University Press), pp.65–89
Zarrilli, P. B. (2004) 'Toward a Phenomenological Model of the Actor's Embodied Modes of Experience', *Theatre Journal* (56: 4), pp.653–66

7
Brecht's *The Threepenny Opera* for the National Theatre: a 3p Opera?

Anthony Meech

Editors' Note: As discussed in our introduction to this volume, the Brecht Estate refused Anthony Meech permission to quote from his and Jeremy Sams' translation of The Threepenny Opera. Consequently, this essay omits the unauthorized material. Rather than not publish this essay, we have instead provided blank spaces with paraphrases in endnotes indicating the substance of each translated segment.

After opening in Canterbury in October 2002, a production of *The Threepenny Opera* (*Die Dreigroschenoper*) by the Education Department of the National Theatre went on a thirteen-venue tour, before returning to the Cottesloe Theatre in February 2003.

The production was directed by Tim Baker and designed by Mark Bailey. The musical supervisor was Steven Edis, and the musical director was Douglas White. The lyrics were translated by Jeremy Sams, and the book by Anthony Meech. This essentially practical study will address the challenges and opportunities faced by us as translators for a contemporary production.

For at least two centuries it has been recognized that the translator has the responsibility to be as well acquainted as possible with the source language, and the cultural and historical environment of the text he or she is translating. In addition, the theatre translator must acquaint him or herself with its genesis and its première, as well as its subsequent production history, in order to gain an awareness of what will be required of the new version when it reaches the stage. Translating Brecht holds a particular fascination. Peggy Ramsay once commented that translation was 'a privileged form of conversation with an author'. In Brecht's case the translator is all too often aware of his barely suppressed laughter at attempts to render effectively his particular and idiosyncratic use of German.

Brecht was such a consummate maker of theatre (a 'Theatermacher') that it is reasonable to assume that whatever appears in the text did, or would, work onstage in the original German. The task of the translator for a British production is then, in one sense, simpler than it might be with, say, Goethe, or more recently Elfriede Jelinek. Brecht's attitude to his own, and others' texts was always a utilitarian one.

The stories from the Berliner Ensemble of the actors' disquiet when the typewriter started clicking during a late rehearsal, confirm that Brecht never regarded the text as immutable holy writ. The text is only one element in the production's attempt to communicate with its audience. For Brecht is primarily a practitioner. And a translator for the theatre must recognize his or her rôle as a member of a company, whose joint aim is the product, the 'production' of a piece of theatre. And, to view that statement the other way round, the translation which reaches the stage is perforce a joint creation. J. Michael Walton (2006, pp.23–4) expresses this concept elegantly: 'For the practitioner there is no ur-text (*sic*), no primary finished version of the original: there is only the text as Ordnance Survey map from which the director and designers [and also in this instance the musical director] fashion the landscape for the players to inhabit.' And the text is never finished. Each production (sometimes each performance) is only a stage on the journey. It is crucial to bear this in mind when working with Brecht texts in particular. It was not only false modesty that led Brecht to publish an edition of his works under the title 'Versuche' (attempts or essays), as Brecht was always intensely aware of the organic, collaborative nature of the production process.

It is something of an irony, then, to encounter the extremely protective attitude of the Brecht estate towards the canon, particularly when so much of what Brecht appropriated and published as his own now appears, thanks to John Fuegi's researches, to be the work of others.

A great many variables fortuitously come together to allow the performance of any play. For a première production to become a national and international success on the scale of *Die Dreigroschenoper* in Berlin is something very rare indeed. The story of its genesis is of a bizarre succession of chance events and happenings, but with significance for anyone intending doing a version. It therefore demands consideration in some detail.

Ernst Aufricht, a somewhat unsuccessful actor, having come into some money, had decided to rent the Theater am Schiffbauerdamm in Berlin and was looking for a new play with which to reopen it on his birthday in August 1928. Having rejected a number of suggestions of works from established playwrights, he went to Schlichter's bar to meet with Brecht.

Brecht spoke at length and enthusiastically of his play *Joe Fleischhacker*, but only when Aufricht and his assistant got up to leave, and Brecht mentioned an adaptation of John Gay's *The Beggars' Opera* (which had enjoyed a successful run in London some eight years before), was his interest aroused. Brecht offered to show him 'some six or seven scenes' the next day. What Brecht had in mind were scenes from Elizabeth Hauptmann's translation of the Gay text into German, which Brecht was intending to adapt into a play to be called *Gesindel* (scum or riff-raff). When Aufricht commissioned the play, Brecht insisted that music for the songs be written by Kurt Weill, who was then just beginning to attract critical acclaim (he had two short operas on in Berlin at the time). Brecht had already established a working relationship with Weill as a composer of songs on the *Mahagonny Songspiel* for the Baden-Baden Festival. Aufricht agreed but, thinking Weill excessively atonal as a composer, had Theo Makeben, the musical director, work up the original Pepusch score, as a fall-back.

After the many twists and turns which the book and lyrics of this adaptation of a translation went through, what hit the stage on the opening night was a composite of an adaptation of Elizabeth Hauptmann's German translation of the original English text by John Gay, a number of songs borrowed more or less verbatim from Klammer's volume of translations of François Villon's ballads, some borrowings from Kipling, some songs which Brecht and Weill had written and set previously, and at least one new song by Brecht ('The Moritat of Mack the Knife'), which was written and set to music overnight to build up the first entrance of Harald Paulsen as Macheath. Indeed, it could be argued that the Moritat is the only element of the book and lyrics which can claim to be truly original and by Brecht.

The changes, additions and subtractions to the text continued until the opening and throughout Brecht's life, although the text generally used as the basis for subsequent performances is the one published in 1931, which includes a number of passages which more clearly define Macheath's character.

Knowledge of this complicated and difficult birth highlights the difficulty a translator faces when approaching the inevitable unevenness of the text of *The Threepenny Opera*, but also served as reassurance to the translators of the 2002 version that sacrilege was not being committed when alterations and emendations were made to the book and lyrics for the National's production.

The work has always inspired mixed responses from critics. Eric Blom, writing in 1931, declared that: 'both words and music of the

Dreigroschenoper look suspiciously as if they had been deliberately written for cretins' (in Hinton, 1990, p.140). He goes on to say (pp.141–2):

> It is a morbid entertainment for those who are sated, not only with all that is sublime, but with everything which is artistically agreeable. For them it has a spice of novelty, though it is quite evidently not a new departure, but a decadent offshoot of socialism, dramatic realism, and utility music.

While intended as a damning comment, the second sentence at least would undoubtedly have delighted Brecht.

At the other extreme of critical response, the adapter for the 1954 Broadway production, Marc Blitzstein said in the sleeve note for the original cast recording: 'I wish I had written *The Threepenny Opera*, but since I merely translated it into English and adapted it for American audiences, I can come right out and say freely that the work is a miracle, a phenomenon, a shining landmark in the history of the international musical theatre.' The piece has had a mixed history on the British stage. As preparation for the first London production at the Royal Court Theatre in 1956, Sam Wannamaker spent eight days with Brecht at the Berliner Ensemble, and had even suggested to Brecht his idea of staging the play as if it were in a club rather than a theatre, to allow closer contact between the actors and the audience. Despite the designs being by Caspar Neher, and the installation of an extended forestage, the production divided the critics. Wannamaker may well have been right to deduce from the reviews that the critics had made up their minds on their responses before the curtain was raised, but his instinct to stage the piece as a cabaret rather than grand opera was surely the right one. This production also had the privilege of a review in the *Daily Herald*, which described it as: 'a show with a violent, shocking difference . . . don't take your maiden aunt' (in Hinton, 1990, p.73).

Two other previous British productions would appear to have come closer to realizing Brecht's intentions of the piece. Annie Castledine's version with Manchester's Contact Theatre in 1994 succeeded because of the company ensemble rehearsal techniques and the theatre's reputation of staging small-scale innovative productions for young audiences. This freedom allowed the company to 'own' the text, while one might have hoped that Castledine's emphatic characterization of the Peachums as bourgeois entrepreneurs, rather than as characters in a Dickensian thieves' kitchen, might have established for all future productions the direct social reference and contemporary relevance of the piece.

While these two productions were both at what might be characterized as experimental stages, consideration must now be given to previous productions of *The Threepenny Opera* at the National Theatre. In 1986 what Stephen Hinton describes as a 'dutiful performance' was staged in the Olivier Theatre, with a 'star-studded cast' and with Tim Curry as Macheath. The production was expensive and resource intensive, the version did not adapt the licensed version of the translated text and the orchestra played every note of Weill's score, but it remained bland and ultimately seems to have missed the point of the piece. It would appear that very little has changed in the main houses at the National Theatre in the last twenty years.

On the other hand Margaret Eddershaw (1996, p.117), in her assessment of the 1982 Michael Bogdanov production of the Caucasian Chalk Circle (also for the Education Department of the National), perfectly encapsulates the aims of the 2002 *Threepenny Opera*:

> with this production [. . .] the National Theatre was openly acknowledging the socio-political and didactic purpose of Brecht's work, and finding a way of allowing an ensemble of actors to make direct contact with an audience. In contrast, the conditions and purposes of their in-house productions of Brecht made such contact difficult, if not impossible to achieve.

It would seem then that an awareness of the socio-political inspiration and an acknowledgement of the didactic nature of the piece, coupled with a close relationship between the cast and the audience might offer the best chance of an effective production.

It could be argued that the theatre has always had the most legitimate claim to be the art of the possible. So many of what are hailed as aesthetic innovations are, in fact, the result of theatrical or financial imperatives. In the case of the 2002 production, the imperatives of small-scale touring led to an inspired decision by the director Tim Baker to cast actor/musicians, with the intention that they would play a somewhat reduced and rearranged score, sustained by a member of the ensemble whose principal rôle would be playing keyboards.

This strategy gave rise to an interesting, and for a discussion of the staging of translated texts, significant problem, which deserves discussion. The logistics of staging the play with musician/actors meant that there were moments when a particular musician could not play because he or she was acting, or could not physically be in the right place. What seemed the obvious answer, to arrange or write some bars of new music, almost prevented the production reaching the stage at all. Until ten days before the start of the tour, the Weill Estate was still refusing permission for the company to use the modified score. Music may be an international language, but its 'translation' for a particular stage environment can be fraught with problems.

One of the aims of the production, a close relationship with its audience, was significantly enhanced by its intimate staging. This was an attempt at incorporating the last of Brecht's unrealized aims for the theatre, 'Die Neue Zuschauerkunst' – the new art of being an audience, or, perhaps the active, or engaged audience. In the 'Lehrstücke' (the teaching plays) with which he was experimenting in the late 1920s and early 1930s, Brecht was trying to involve his audiences as equal partners in his productions, in order that they would recognize the relevance of the stage action to their own situation. To bring this about, he wanted his actors to enter into a different, more equal contract with their audiences.

In Tim Baker's *The Threepenny Opera* the interpolated, improvised prologue, in which the actors (out of character) asked questions of the audience, immediately established precisely this direct contact, while reinforcing in the minds of the audience the ensemble nature of the company. This prologue was substituted for the one, which is commonly used – the one delivered by Gerald Price in the Decca 1956 Broadway original cast recording – (although there is no prologue in the 'Stücke' edition of the original German):

> You are about to hear an opera for beggars. Since this opera was conceived with a splendour only a beggar could imagine, and since it had to be so cheap, that even a beggar could afford it, it's called *The Threepenny Opera*.[1]

The 2002 Prologue differed depending on the performance venue and the audience. An excerpt from the very end of the prologue from one of the nights at the Cottesloe will give a flavour of this improvised interpolation, as spoken by David Rubin:

> and we'd like you to imagine that this group of beggars is angered by the situation we find ourselves in, so we've written an opera. And so that we can produce our opera, we've broken in to the National Theatre, and we've stolen some props and some costumes, and some musical instruments.

Margaret Eddershaw (1996, p.118) also says in the chapter quoted earlier:

> To all intents and purposes Brecht's plays have been widely accepted into British theatre, but his explosive, theatrical power and stimulus to social change were largely defused.

The Threepenny Opera has now become a classic, and has (mostly in larger scale productions it would seem) acquired that supreme impotence which Brecht himself despised in the classic. How can a translator restore the original vitality and bite of such a familiar piece? How can a contemporary interpreter reinvigorate a text, and make it speak to a modern audience? The crudely painted message 'Homeless and Hungry' on the backcloth, as well as the reference to the local homeless living on the streets ('in Covent Garden' at the Cottesloe) served forcibly to bring the performance's message home to its audience. The message was all the more readily received by the audience, who, through the prologue, had established a genuine and close rapport with the ensemble. The reference to the plight of the homeless was recapitulated in the third Threepenny Finale at the Cottesloe, when Harvey Virdi (as Mrs Peachum) and David Rubin (as Peachum) sang:[2]

When considering the respective advantages of 'domestication' and 'foreignization' in their simplest forms the translator does not as a rule face the peculiar problems posed by *The Threepenny Opera*, which is set in the country of the target language. The question facing the translator is how to translate English into English? Is it possible to achieve an impact on an English-speaking audience parallel to that English ambience and place names have on a German audience? There is a challenge for a translator writing for an English audience in *Mahagonny*, where the Alabama Song is written entirely in a sort of English, and the opera set in what his audience are likely to be predisposed to accept as a fanciful interpretation of America. But the challenge is more acute in *The Threepenny Opera*, as the piece is set in Brecht's fanciful recreation of an England which never existed, except perhaps in Hammer Horror films. In a similar way in which the Chicago of *In the Jungle of Cities* and *Arturo Ui* draws more on the cinema and detective comics, the London of *The Threepenny Opera* shows little evidence of any research into nineteenth-century London and its environs.

It does, however, create a unique world into which any production must invite its audience, and it is part of the translator's role to assist the director by defining the contours of an equivalent world, peculiar to the specific production. This raises the question of how German *The Threepenny Opera* is. Despite its setting in a sort of London, the play remains essentially foreign, though not necessarily German. Mark Bailey, the designer, made the useful suggestion that the text might be regarded as European, rather than either German or English.

Similarly, the play was originally set somewhere in the late nineteenth century, but the queen is about to celebrate her coronation, which would place the action in 1837. It is simply too easy for a British designer to set the play in the swirling mist that always seems to cling around be-caped 'Bobbies' in that image of London, and allow the audience to relax and engage with it with the same relish as they would with a Hammer Horror film, or perhaps the Rocky Horror Show.

It was vital for the success of this staging to create a discrete yet coherent world for the production to inhabit, one which the audience would be prepared to enter, or 'buy in to' for the piece to have its desired effect. The decision was made very early on to set the production in no specific time but more or less contemporaneously. (The change from 'The Queen's Coronation' to 'The Queen's Jubilee' seemed obvious.) Minor changes were made elsewhere. Amounts of money were simply updated to make them credible, much of the slang was sharpened (introducing, for example, 'You're screwed' and 'Did you shag her'). These numerous, but in themselves relatively minor changes helped to create an edgy, flexible and responsive linguistic environment, which allowed for what were regarded as the vital amendments and interpolations in the script.

The world which Mark Bailey created in his design, while recognizable by these contemporary references, was a nonetheless consciously strange and foreign place, where prisoners could be locked up behind bars with enormous keys, despite the gang declaring that they had stolen Sir Elton John's harpsichord. It was precisely these disjunctures which served to keep the audience alert to the inserted anachronistic references. When Tiger Brown complains that Macheath has perpetrated yet another break-in to hold his wedding in a 'borrowed' cellar (not a stable) Mack retorts: 'It's alright it's all above board, it belongs to the Prime Minister's wife, I helped her to buy it.' This contemporary reference to Cherie Blair's purchase, in somewhat dubious circumstances, of a flat for her son was immediately pounced on by audiences. But the interpolation best remembered by audiences resulted from a fortunate news story, which broke during rehearsal, which allowed James Lailey, as Tiger Brown, entering as the preposterous Mounted Messenger with his mock-Handelian recitative announcing the outrageous royal pardon at the end of Act Three, to drop the lines: 'On the occasion of her coronation, the Queen commands that Captain Macheath be immediately set free' and substitute the following:[3]

A number of members of the audience commented that the whole opera had been made more political, when in fact the changes which had been made, were few and far between.

Potentially the greatest challenges for a translator are those mentions of the British Raj in the most directly historical of the songs which come from a fantasy of Empire (where in *Man is Man* the soldiers have names like Jeriah Jip and Blody (*sic*) Five and all drink whisky, as in 'The Cannon Song'). But the line 'From the Cape to Couch Behar' has little or no current resonance for us, and serves only to historicize the song in the wrong way, allowing the audience simply to enjoy the catchy and familiar tune. To prevent this Jeremy Sams wrote joltingly contemporary lyrics for the number, which included the chorus:[4]

Indeed, Tim Baker recently suggested that a production now would need to introduce additional references to Afghanistan as well as Iraq to maintain the shock value of 'The Cannon Song'.

These radical alterations to the lyrics raise the more general relationship of the words of *The Threepenny Opera* and its music. When the music was first played in 1928 it was arrestingly, even shockingly, modern. Indeed, what staggered observers, such as Adorno, was that the music could be both modern and popular. His answer, of course, was to suggest that the audiences did not really understand the music. Nowadays the music inspires nostalgia, rather than outrage, and a familiarity which allows us to wallow rather than listen. But the expedient of updating the score, as was possible with the book, would never receive the approval of the Weill estate. Instead, Jeremy Sams was allowed much greater licence in translating the lyrics than was thought appropriate with the book. If the music no longer made the audience sit up and take notice, the lyrics often did.

In this way the translators attempted to discharge that responsibility translators of any Brecht text must recognize, of incorporating the 'social Gestus' of the original in their version. If the 'Gestus' is not apparent in the text, it is futile and dishonest for an actor to try artificially to imagine his or her 'Gestus' and to try somehow to inject it into a speech. During my work with them as dramaturg, the cast were keen to talk about 'Verfremdung' (Alienation) and 'Gestus', and how they might introduce these into their performances. Before meeting with the cast I had suggested that they should listen to the words they were saying, and, in particular, singing, and stress the fundamental immorality and nastiness of the characters – above all of Macheath. This would be in sharp contrast with what has become the accepted reading of the character of Macheath as a cheeky crook, perhaps best exemplified by the smooth, 'easy listening' approach adopted by Bobby Darin in his version of the 'Moritat', with its interpolated references to 'old Lotte Lenya and Miss Lucy Brown'.[5] Jeremy Sams, by contrast, ended his version of the 'Moritat' with the couplet:[6]

The suggestion was made that the cast follow Brecht's own advice and find all the theory and 'Brechtian acting' they would need in the text itself. A frequent problem with English actors playing Brecht is that they have a sense that there is something called Brechtian acting, which critics with experience of the Berliner Ensemble for instance would instantly recognize (if they could achieve it) and to which a dramaturg should have the magic key. When I watched them in rehearsal, I was tempted to repeat Brecht's famous response, when Lotte Lenya asked him if she was being 'epic' enough. 'Anything you do is epic enough for me'. There can surely be no better concrete example of a *Verfremdungseffekt* than the sight of Polly playing her trombone while her Mother was berating her for marrying Macheath. It would be hard to find a better example of gestic singing than Elizabeth Marsh (who was as surprised to find she was already doing gestic singing as Molière's M. Jourdain was to discover that he had been speaking prose all his life). She sat astride a chair in the brothel in scene five to deliver her caustic, world-weary version of the Pirate Jenny song (Figure 7.1). In her singing of this familiar number, she was guided towards finding the 'Gestus' for her performance by the Jeremy Sams' re-invigorated lyrics:[7]

Her provocative position astride the chair downstage, directly confronting the audience, was subverted by her blank facial expression, which only once nearly became a smile on the lines:[8]

Figure 7.1: Elizabeth Marsh singing 'Pirate Jenny's Song' in *The Threepenny Opera* (*Die Dreigroschenoper*) National Theatre, London, February 2003

This performance stands in stark contrast to the 1954 sanitized version of the song by Lotte Lenya (Jenny in the 1928 première production) on the Decca recording of the Broadway production, showing how thoroughly she had been Americanized.[9]

The transfer of a text via a translation to the stage inevitably involves a series of compromises. The playwright makes a text, which, before it can reach its target audience must pass through the hands of a succession of practitioners – people whose practice will develop, interpret, even sometimes radically alter, the original.

At the 2007 Mülheimer Theatertage there was much heated debate about what the Germans call 'Texttreue' – faithfulness to the text. German directors have traditionally assumed they have *carte blanche* to alter play-texts in order to make on stage the production they envisaged. In 'Regietheater' (Director's Theatre) the text serves as a bolt of fabric to be cut and stitched together by the director. In Britain the situation appears very different. Contracts are signed which require the director to seek permission before making any changes to a performance text. This can severely restrict an imaginative director, who may have problems dealing with an author who sees his or her work as holy writ, but it can offer exciting opportunities for the director of a translation, who can use the requirement to consult with the translator to enlist his or her help to wrest the true meaning from a particular element of the text.

It is, of course, true that a rôle of the translator is to act as guardian of the text, but, if the translator is also a dramaturg, consultation with him or her can offer the director that 'elbow room' in which the staging can evolve. Tim Baker also talks of a good text as being one which one can 'lean on' (Baker 2007). It must be the responsibility of the stage translator to help his or her director to sense how far he or she can lean in creating a vibrant living version for his or her audiences. Above all a translator/dramaturg must be prepared to sanction cuts to his or her script when a passage manifestly does not work in that particular staging.

It is a truism that, while masterpieces are immortal, translations have a limited life-span. It is the very evanescence of a stage translation, which lends it its immediacy and vitality. The English, after all, might be said to have only one Shakespeare, while, in translation, there are as many Shakespeares as translators and casts to perform them. It may in fact be true now, as Tim Baker suggested, that the country where it is likely to prove most difficult to realize *The Threepenny Opera* successfully is in fact Germany.

In a recent interview Tim Baker said that he would love to do *The Threepenny Opera* again, using the same translators, but this time spending more time with them around a table working through the text – to allow himself more 'elbow room' as a director (Baker 2007). After his unique experiment in 'gestic' translation in America with Charles Laughton on *The Life of Galileo*, I am sure Brecht would approve.

Notes

1. All quotations from the 2002 National Theatre text are taken from the unpublished prompt copy.
2. Mr and Mrs Peachum sang a version close to the wording of the original German, but used as examples the homeless in Covent Garden.

3. Instead of the Queen's coronation, the Mounted Messenger referred to the then current news story of the Queen fortuitously remembering a conversation (which led to the collapse of the trial of Paul Burrell).
4. In the Cannon Song there were references to Bosnia, Goose Green and 'Squaddies'.
5. Bobby Darin's version of 'Mack the Knife' is available in numerous recordings including *Beyond the Sea: the Very Best of Bobby Darin* (Warner Music Group) and the DVD *Bobby Darin*, 2005, Frechen, Delta Music GmbH.
6. The couplet refers to Macheath as a rapist and a sadist, and the fact that he has not been caught.
7. In Pirate Jenny's song Jeremy Sams used slang words like 'punters', 'skivvy', 'whore' and expressions like 'haven't got a clue'.
8. The translation of 'sag ich "Hoppla!"' is 'I say: "Cool!".'
9. The Lotte Leyna version of Pirate Jenny's Song referred to can be found on the Decca Broadway Original Cast Album recording of the 1954 Carmen Capalbo and Stanley Chase stage production of *The Threepenny Opera*.

References

Baker, T. (01 March 2007) Unpublished interview
Eddershaw, M. (1996) *Performing Brecht* (London: Routledge)
Hinton, S. (1990) *The Threepenny Opera* (Cambridge: Cambridge University Press)
Walton, J. M. (2006) *Found in Translation* (Cambridge: Cambridge University Press)

8
The Translator as Cultural Promoter: or how Renato Gabrielli's *Qualcosa Trilla* went on the Road as *Mobile Thriller*

Margaret Rose and Cristina Marinetti[1]

While scholarship in translation studies makes frequent reference to theatre translators as cultural mediators (Bassnett 1998, Aaltonen 2000), the term is usually intended in a textual sense, indicating the function that translators, in their role as intermediaries between the source and target text, play in negotiating meaning between cultures. In this article, which is based on Rose's experience in the translation and British production of Renato Gabrielli's *Qualcosa Trilla*, we suggest that the specific contexts of practice of this theatre translation project point to a much wider set of roles covered by the translator/cultural mediator. These extend far beyond the delivery of the text in the target language and include: creating a market for the translated plays (by raising funds for production, organizing festivals and other events) as well as following the text of the play through production, rehearsal and site specific performance. We also suggest, in a vein that supports the overall methodological stance of this collection, that studying specific contexts of practice of translation phenomena through a combination of textual analysis, practitioner's account and critical reflection, can offer innovative insights into our understanding of translation and can be a valuable counterpart to theoretical considerations based on descriptive analysis.

From cultural mediator to cultural promoter

There is a significant difference in cultural policies between Italy and the UK in their reception of foreign language plays. While a large number of contemporary British plays are imported to Italy every year, the number of Italian plays arriving in Britain is exiguous. Italian theatre seasons regularly feature many foreign plays, with a high percentage of British and Irish works. In the 2008–2009 theatre season, Shaw's *Mrs Warren's Profession*, Ravenhill's *Shoot/Find the Treasure/Repeat*, Beckett's *Endgame* and *Waiting for Godot*, Berkoff's *Kvetch*, as well as *Living Things*, a retrospective on Harold Pinter (*Sipario* 2009), to name just a few were produced. In the UK the story is much bleaker. Even if most theatergoers have heard of Carlo Goldoni, Eduardo De Filippo, Luigi Pirandello and maybe Dario Fo and Franca Rame and may even have seen some Italian plays in performance, they are unlikely to be aware of a host of young and not so young contemporary writers who have emerged in Italy in recent decades (Puppa 2006).

These differences in the value and cultural currency of foreign theatre in Britain and Italy have a considerable impact on the specific contexts in which theatre translators operate and in the degree of involvement of the translator in the phases of production, rehearsal and staging of the translated play. The reliance of many established theatres in Italy on productions of translated plays has produced a context of practice where new foreign plays are a sought after cultural commodity and translation and translators are an integral part of the production process. In Britain, where established theatres have tended to favour new plays by British authors and classics over foreign plays,[2] translators of contemporary foreign plays find themselves on the edge of the theatre market. This is where the translator, if he or she wants to see their translations performed in the UK, can move from the position of cultural mediator to that of cultural promoter and actively seek to create a market for the foreign play, working alongside producers, directors and actors as an integral part of a creative project.

What follows is an exploration of the processes (textual, contextual and performative) involved in Rose's translation/co-adaptation of the one-man play *Qualcosa trilla* by Italian playwright Renato Gabrielli.[3] The discussion covering the various metamorphoses of *Mobile Thriller* (Gabrielli unpublished, and Rose 2004)[4] between 2003 and 2005 takes in the difficulties and issues encountered in the translation/adaptation process and its critical reception. In particular we will be considering the translator's role in a process which included translation, adaptation and staging and a site-specific production and which will help us trace the trajectory of the script from source text to target production. In the creative team, Rose worked as translator and co-adapter and cultural mediator but also, as we shall see, as a member of the creative team, she was involved in creative decisions all the way through the adaptation and staging process. For the translation she worked mainly alone, except for conversations with the author, a situation which changed when the work on the adaptation started. At that stage the translator became part of a team composed of writer, director, producer and actor, resulting in an adaptation which includes many suggestions and ideas deriving from group discussions.

Rose's involvement with the translation, staging and production of *Mobile Thriller* provides an example of this broader function of cultural mediation which is not much talked about in the literature on theatre translation but seems vital to the survival and development of a fruitful exchange of ideas, texts and modes of performance between theatre systems. The methodological choice of combining practitioner's account with critical reflection involves occasional shifts in this essay from impersonal academic discourse to Rose's first person account.

'Creating the market' for my translation: the *Scambiare* Project

In 2003 I had the idea of setting up an Italian-Scottish arts festival in the hope that this would arouse interest among theatre practitioners and audiences in Scotland. Together with Hugh Hodgart, Director of Acting at Glasgow's Royal Scottish Academy of Music and Drama (RSAMD), I created *Scambiare* (*Exchange*), a wide ranging cultural programme involving Italy and Scotland. On our agenda were drama, film, music and a seminar, 'Strange Behaviour' (the latter organized by the Glasgow-based theatre company, Suspect Culture). We were fortunate in that RSAMD, Suspect Culture, Strathclyde and Milan Universities, CCA (Glasgow's Centre for Contemporary Arts), the Italian Cultural Institute in Edinburgh and Outis (The Italian National Playwriting Centre) were persuaded to join forces to provide a minimum of funding, facilities, publicity and travel grants. However, following the first festival, which I had envisaged as an annual event, taking place on alternative years in Glasgow and Milan, with heavy traffic both ways!, no further funding materialized and Hugh Hodgart and I were forced to call it a day.

Prior to *Scambiare* I did some preliminary research to decide which plays to include in the theatre section and came up with six very different plays, thematically and stylistically speaking, which would hopefully challenge Scottish audiences. A team of translators set about translating the plays. At RSAMD Hugh Hodgart set up a workshop, and trainee directors and actors were organized in small groups to work on one of the translated plays. Our aim was to stage an advanced rehearsed reading of an excerpt from each play for an audience made up of students, the general public, and a carefully selected group of theatre people and critics. During the festival the Italian playwrights visited Glasgow to see their work in translation, take part in a post-performance discussion and make contact with their translator and theatre people in Scotland.

In what was still a rather heady post-devolution climate, critics tended to point to *Scambiare* as an important link fostering cultural exchange between the two countries. The title of Mark Brown's article in *The Scotsman* (Brown 2003) read, 'Three Cheers for the New Alliance', with this critic positioning *Scambiare* in the aftermath of the turbulent events linking Italy and Scotland during World War Two in contrast to the present-day success story of many Italian immigrants in Scotland. Despite this enthusiasm, only *Mobile Thriller* enjoyed a subsequent life on the road.

The play: a tale of suicide and death

This one-man play has as its protagonist, Massimo, a wealthy Milanese businessman. At around forty he appears to have achieved everything; a well paid job and a successful career, the latest gadgetry and trappings of modern life, a superb car and a luxury home, a wife and child. Instead he is desperate, eaten up with self-loathing and has decided to end his life. The play re-enacts his car journey through Milan and out into the suburbs, where Vincenzi, a hired killer, is waiting to shoot him. Massimo talks on his mobile to his wife, child, killer and his best friend Carlo. On occasions he speaks into a Dictaphone, recording a message he intends leaving for an unnamed woman, a real life, or more probably, an imaginary lover, whom he is passionate about. He seamlessly switches from one addressee to another and back again, in a deeply fragmented monologue, as if zapping between television channels. Linguistically, then, the play reflects the fragmentation in our daily lives, both psychologically and linguistically, partially caused by mobile phones and remote controls.

Through Massimo's sad plight, we are made familiar with Milan's underbelly, where unbridled materialism is rampant, rather than the more widespread cultural narrative concerning Milan as a glitzy world of designer clothes and fashion queens. *Mobile Thriller* struck me, moreover, as a powerful modern-day tragedy, with a great deal to say about the dark side of life in any big city. Massimo's depression and suicide are certainly not confined to Italy, but emblematic of much modern day angst, caused by interpersonal relationships distorted by new technologies and the alienation of big city living. These ingredients can easily be appreciated by audiences in big cities throughout the world.

Comparing texts: the play, the English translation and the Fringe adaptation

When I translated the play for the *Scambiare* readings in Glasgow, an adaptation transferring *Mobile Thriller* to Scotland seemed unnecessary. I started by making a rather free translation of the play's title,[5] foregrounding the centrality of a mobile phone, rather than just an object which rings as suggested by the Italian title. In the English title I wished to highlight the genre of the play which might be defined as a theatrical thriller. The precise reasons for Massimo wanting to die, the identity of his killer and the enigmatic figure of his lover, Piggy, remain shrouded in mystery. As in a thriller, tension mounts as audience members gradually come to realize that Massimo is intent on suicide.

My aim was to re-create in the target language the different linguistic registers in the source text. I endeavored to ensure that the linguistic shifts, often rapid and surprising in the source text, remained in the target text. They sustain the dramatic tensions and rhythms as well as producing comic effects at times. One of these registers is the colloquial Italian spoken in Lombardy, and particularly in Milan. This is found whenever Massimo talks to his wife and especially to his young son, Gustavo. In the following extracts Massimo switches rapidly from talking to his wife and his son on the mobile to the lover in the imagined conversation/letter, which surprises audience members and produces laughter:[6]

Massimo: Me lo passi?
Porca.
Gustavino, fai il bravo con la mamma. D'accordo? Me lo prometti, che mangi la carne?
Porca.
No, papà torna tardi. Mi ripassi la mamma? Pronto. Pronto.
Scusa se ti chiamo porca mentre parlo con mia moglie e mio figlio. Probabilmente non lo sei e in fondo neppure ci conosciamo, non è questo il punto. Ma tu capisci quant'è essenziale per me poterlo fare in questi momenti. Nei momenti di tepore familiare, pensarti porca, ah, angelo mio, capisci?

(T1)   Put him on
You dirty bitch

| | Be a good boy for your mummy. Understand? Eat all your meat up. Promise? every teeny bit. You dirty bitch. No your dad'll be back late. Put mummy on again. Hello? Hello? Sorry if I call you a dirty bitch when I'm talking to my wife and son. You're probably not dirty at all, after all we don't know each other that well, but that's not the issue. Still you can understand why I need to call you names at times like this, the moment I feel the warmth of family ties, that's when I feel you are a very dirty bitch, my angel. You do see don't you? (p.43) |

(T2)   Will you put him on?
You dirty Piggy.
You being a good boy for mummy, and eaten all your dinner? Promise me? Eat up all your meat, come on? Make you big and strong.
Dirty Piggy.
No, dad'll be back late. Put mum on again. Hello? Hello? Sorry if I call you a dirty piggy when I'm talking to my wife and son. You're probably not dirty at all, after all we don't know each other that well, but that's not the issue. Still you can understand why I need to call you names at times like this, the moment I feel the warmth of family ties, that's when I feel you're a real dirty piggy, my angel, you do see, don't you?

The key to this passage and to the characterization of Massimo is his ability to move from one conversation addressee to the next with great skill in a matter of seconds whilst remaining believable in all his roles. In the Italian this is rendered by linguistic choices which perfectly suit the type of relationship he has with each interlocutor and this is particularly evident when Massimo addresses his son. Here he uses the endearing diminutive form for the son's name (Gustav-ino instead of Gustavo) as well as a simplified, childlike sentence structure ('Me lo prometti, che mangi la carne' instead of the more formal and more correct form 'Mi prometti di mangiare la carne'). This makes him sound like a very believable family man, father and husband. A radical change in register occurs whenever Massimo talks to his lover, an unidentified woman, whose pet name is 'Porca', literally 'a sow', or in slang, 'a slut'. Originally 'porca' was translated as 'dirty bitch' for the initial readings (T1). However, after conversations with the author, who described his choice of 'porca' not so much as an ironic insult but as a proper pet name suggesting erotic and playful innuendos, 'porca' was rendered as 'piggy' and more often 'dirty piggy' during the adaptation process (T2).

Other changes brought about by the collaboration of translator, original author and actors during the adaptation and rehearsal process have to do with finding a 'speakable', colloquial equivalent to Massimo's way of addressing his son. While the solution in (T1) is correct and adequate for a father son relationship it retains a level of formality that does not bring to life Massimo's role as a family man and father. (T2), which was developed during rehearsal, on the other hand offers a solution which is fluent, colloquial and speakable, especially through the replacement of the original imperative 'be a good boy for your mummy' with the more natural and colloquial present continuous with the elision of the auxiliary '[are] you being a good boy for mummy'. This, and the other expression typical of a way of talking to (talk directed at) children ('Make you big and strong') brings to life the relationship between Massimo and his son in English as a believable father/son relationship. This moment of family life is very important because in the play it acts as a counterpoint to Massimo's extramarital relationship with Porca and shows two sides of his personality which are clearly conflictual. In the final passage of this extract, where Massimo seeks to explain why he has taken to calling his lover 'porca' – to create distance between her and 'the warmth of family ties' – , it seems, anachronistically, that Massimo could find love, warmth and understanding only with her ('my angel'), only outside the stifling burden of family responsibilities. And the nickname 'porca' becomes a way of exorcizing, through language, the danger of remaining trapped within established relationships.

In the sequences which Massimo addresses to 'Porca', in the Italian original, the language waxes lyrical and is sometimes aggressive, characterized by complex syntax and a high-flown style, which again creates a stark contrast with the prosaic and colloquial language of his interactions with his son and wife, which becomes steadily more prosaic and colloquial as the play goes on:

Massimo: Sì? Ancora tu? Gustavo, ora basta coi capricci! Se no, vuol dire che non sei grande ma ancora piccino, Nanetto, granchietto, rospetto, perché piangi, passami la mamma piuttosto. Ascolta, vuoi tenerlo lontano dal telefono, per piacere?

(T1)		Ah it's you again. Listen kid I've had enough. You're behaving like a baby not a big boy. A babe in arms, a wee baby, babino, bubino. Now stop whining, you little puppy wuffy, wuf, wuf, my one and only puppy wuffy, put your mum on please. (p.43)
(T2)		Ah, it's you again. Listen kid, I've had enough! You're behaving like a baby not a big boy. A babe in arms, a wee wee baby. *(He starts singing softly, towards the end a touch of anger in his voice.)* Hush a bye baby on the tree top / When the wind blows the cradle will rock / When the bough breaks the cradle will fall / and down will come cradle, baby . . . and all.
		Not again. I'm fed up with your whinging you little pissy pants, put your mummy on please. Listen, can't you play with him, keep him off the phone.

In (T1) and (T2), rather than a literal translation of 'Nanetto, granchietto, rospetto' ('little dwarf, little crab, little toad'), I invented an expression a father might yell at a small child in a British context ('little puppy wuffy, wuf, wuf, my one and only puppy wuffy' and 'you little pissy pants'). Noticing how in the original the degree of anger in Massimo's speech increases rapidly at this point, for the adaptation I opted for the more aggressive and less playful solution 'little pissy pants' (T2). Also, in (T2) I inserted a fragment from the lullaby, 'Hush a bye baby', to reinforce the creation of those 'moments of family warmth' which, as we discussed with the previous example, play a very important role in the development of Massimo's character. Another reason behind the choice of a song was linked to what happened in rehearsal, where I discovered that David Walshe, the actor playing Massimo, had a beautiful Irish singing voice, which might well have a powerful impact on the audience.

Another element of the play that proved challenging on the page was the humour which, in Gabrielli, function very much on a stand-up model. Later in the play Carlo, Massimo's old friend from his school days, calls him on his mobile to tell him a tragicomic story aimed at pulling Massimo out of his depression. Massimo decides to retell Carlo's story for the audience, using colloquial Italian, containing a smattering of dialect.

Massimo:	Il Pelizza è un personaggio inventato da Carlo, ma reale, un imbecille medio di questa città. L'abitante modello di

questa città. L'ansia e la spocchia del produrre, il cosiddetto umorismo bonario del *Ué pirla*, un'etica da quattro soldi di borghesia illuminata al neon: tutto questo è il Pelizza. Ti piace? Ascolta: 'Ué, senti qua: il Pelizza incontra il Biraghi al coffee break e mentre buttano giù due caffè in venti secondi il Biraghi gli fa: ué. Ué, risponde il Pelizza, allora il Biraghi gli racconta che hanno inventato una nuova macchineria al top dell'avanguardia tecnologica, il Nanetto Salvatempo, che però non si può ancora comprare, perché è fuorilegge. Ma va là, pirla, dice il Pelizza, poi di nascosto si fionda al mercato nero che c'è sul marciapiedi di fronte all'azienda, e lì ci trova subito un mafioso con gli occhiali a specchio e il Nanetto sottobraccio.

(T1) Pelizza's a character Carlo's invented, but very real, one of the million idiots you find hanging out round these parts. A model Milanese citizen, a workaholic and a real charmer, whose opening gambit is invariably, 'How you doin', fuck face?' He's always ready to take any sod for a ride, a middle-class guy for whom the word 'ethics' means shit all, seeing that his mind flashes on and off like a neon light at the mere mention of money. So here's his latest. Pelizza meets Biraghi, another blue blood Milanese, in the coffee break and while they're gulping down two coffees in twenty seconds Biraghi goes: 'What?' 'What?' replies Pelizza, and Biraghi continues: 'They've just invented a fucking incredible gadget called the time saving Mini Dwarf but it's still not on sale and it's illegal.' 'I'm having one this second', retorts Pelizza, dashing out of the factory. On the other side of the road there's a street market, selling masses of dodgy stuff and Pelizza spies a mafia type, with the predictable dark glasses, big moustache, swaggering along with a Mini Dwarf tucked under his arm. (p.43)

The humour in the original arises from its reliance on shared knowledge and enjoyment in the exaggeration of dialect features ('ue' 'ue' and the definite article before proper names 'Il Pellizza'). The fictitious Milanese businessman immediately rings a bell with audience members both as a possible reference to prime minister Berlusconi (a Milanese businessman by birth, often referred to in the satirical press as 'il nano malefico' 'the evil dwarf') and as a character immediately recognizable as a stereotypical Milanese businessman whose vast wealth is displayed in ostentatious clothes and cars. In general I translated this story into very colloquial English with some expletives. For example, I rendered the Northern Italian 'Ué pirla' (dictionary definition, 'fucker', 'dick' or 'prick') rather loosely as 'How you doin', fuck face?' in a bid to find a similarly colourful expression in colloquial English. However, such examples of stand-up comedy are very rarely translated satisfactorily on the page. The degree of local colour and the amount they rely on building an intimate relationship with the audience, make them very bound to the site of performance. We shall see later in the essay how stand-up comedian Greg McHugh rewrote this passage successfully for a Scottish audience at the 2004 Edinburgh Festival.

Who decided to stage *Mobile Thriller* and why?

Carrie Cracknell, a student on the MA course in directing at RSAMD, directed the reading of *Mobile Thriller* for the *Scambiare* event. Carrie was immediately drawn to the play for its blend of the surreal and real, its magical love poetry and cynicism characterizing Massimo's world. Very rarely, she maintained in a programme note for the Tron production, does such a combination exist in the mainly naturalistic plays being written and produced in contemporary Britain (Cracknell, 2004).[7] Moreover, Hush Productions, Carrie's newly fledged company, had on its agenda the production of contemporary British plays as well as new writing from abroad; Renato's play struck Carrie as a fine opportunity to start this process (Hush Productions website).[8] Irish actor David Walshe, in the role of Massimo, managed to capture the play's contrasts and quick shifts superbly, a rendition which suggested the play's potential in a full scale production.

Following *Scambiare*, Carrie Cracknell and Hush producer James Erskine managed to secure a run for *Mobile Thriller* at Glasgow's Tron theatre, together with *Vocation* and *The Number Ninety's Child*, two other one-act plays by Renato Gabrielli, dealing with fantasy, suicide and the trials of urban living. The three plays, marketed under the collective title of *Death and the City*, received excellent notices. Susan Mansfield's review (Mansfield 2004) in particular led me to believe that the translation had worked well:

> Though they [*Vocation* and *Mobile Thriller*] contain flights of fancy, these visceral pieces of theatre remain rooted in the detail of the everyday. Milan comes alive, a workaholic city momentarily infused

with Vittorio's magic, a dark predator trapping Massimo in its maze of streets. Gabrielli's texts play with the boundaries between fantasy and reality through the minds of two characters who have forgotten how to separate the two.

Mobile Thriller's stage life did not end here. Philip Howard, artistic director of Traverse, Scotland's new writing theatre, came to see the trilogy and decided to include *Mobile Thriller* in his Edinburgh Festival season. The play had won an unexpected ticket to the Fringe!

The metamorphosis continues: *Mobile* 'on the road' at the Fringe

With this exciting prospect in mind, we set up a creative team made up of Renato, Carrie, David, James and I. It was at this point that Carrie decided to transfer *Mobile Thriller* from a theatre stage to a moving car. Massimo would set out on an hour's drive along a carefully planned route from the heart of Edinburgh into the suburbs, with three audience members in the back seat of the car.

The Edinburgh Fringe thrives on novelty and the number of site specific productions has increased in recent years. We were convinced that in a site specific production the acculturation of the play would mean it could communicate more powerfully and with greater immediacy with UK audiences. In the site specific adaptation, the landscape Massimo sees as he travels through the city evoked in the source text through language could be described more precisely, thanks to the invention of new lines. My involvement in the adaptation process acted like a shot of adrenalin, since I was able to bounce my ideas off the rest of the creative team. Especially hearing David Walshe reading the lines, and later performing them, made me realize what was working in the translation and the parts which needed further attention.

As the process of adaptation began, the first major decision involved the transfer of cultural markers. Names, places and other cultural markers were switched. Massimo becomes Donal, Carlo, Stevie, Gustavino, Jamie, Vincenzi, Dunbar, and Porca, Piggie. Donal takes up residence in Balerno, a wealthy district of Edinburgh, and his materialistic Scottish wife shops for clothes at Harvey Nichols. Stevie's story includes a visit to an Edinburgh street market rather than the Italian market indicated in the source text.

Given our innovative 'theatre venue', the introduction of a second character, a chauffeur, who would drive the vehicle and engage in silent action with Donal, seemed indispensable. This left Donal free to handle his mobile phone and Dictaphone. He could also engage in some eye contact with the audience through the tiny mirror in the sunshade and wing mirror and turn round occasionally to peer through the back window.

A soundtrack of US and British songs, like Roy Orbison's 'In Dreams' and Unkle's 'Rabbit in the Headlights', selected by Carrie, created an emotionally charged atmosphere as the car sped through the city that kept audience members spellbound. The English words of the songs signalled the distance the action had travelled from Italy. I added the nursery rhyme, 'Hush a bye baby' towards the end of the play. This dark tragic lullaby seemed to me to capture Donal's mood as he prepares himself never to see his little boy again.

With Renato, I travelled in the back of the car several times along the route, writing new lines such as Donal's instructions to his chauffeur: 'Go across the Meadows' and later, 'Head for the ring road . . . '. These had to describe what audience members would actually see through the car windows on their travels: 'At this very second we're coming into a medley of small shops, pubs, takeaways, hairdressers. It's an ugly bit of the city but the most real . . . '

As was mentioned earlier, the Edinburgh writer and stand up comedian, Greg McHugh, was commissioned to rework Stevie's allegorical story about consumer-crazy Milan. This tale in colloquial, working-class Italian is extremely funny in the source text. We all felt that Stevie's language should point to the fact that both he and the characters in the story hail from Edinburgh. The original setting in a street market in Milan was turned into Edinburgh's vibrant Ingilston market. The familiarity of place and characters would hopefully allow spectators to vividly imagine the place Stevie is describing and the criminal types who go there. The degree of familiarity was enhanced by the many idioms and colloquialisms typical of Edinburgh street language. Greg McHugh's version of Carlo's story for the 2004 Edinburgh production reads:

> Donal: Eh two suits from Edinburgh (*finding his thoughts*) Right, whit hud happened wis . . . that these two (*puts on quasi Edinburgh accent*), Murray and Cameron had met in a bar in Edinburgh, like to talk business . . . boring shite, you know what these fannies are like. The mere mention of money sets their minds flashing on and off like neon lights! Cameron is a model Edinburgh citizen, kind of guy that would take any fucker for a ride, a workaholic and a real charmer, whose opening gambit is invariably, 'How you doin, fuck

face? ... Right anyway, so after a bit, out of the blue Murray says: 'They've just invented an incredible gadget, called the time saving Mini Dwarf but it's still not on sale and illegal.' Cameron, almost creaming himself says: 'I'm having one this second!!!!' Murray then tells him that he needs to git his arse doon tae Ingilston market and to head to the final stall, to the guy who looks like a Skinny Don Corleone, wi Ginger hair and the thickest dark glasses you've ever seen ...

The re-writer, using a formula typical of stand-up comedy, wastes no time in introducing the two main characters, Cameron and Murray. Nouns like 'suits' and 'fannies' and verbs like 'creaming' make the piece more colloquial than my translation. The Scots idiom, 'git his arse doon tae Ingilston market' is funnier and coarser than 'get down to Ingilston market'.

Prior to the start of the 2004 Fringe Festival, *Mobile Thriller* had been publicized as the 'smallest show on the Fringe', targeting an 'audi-ence' of three (the car was a state-of-the art German Audi A8) and was sold out before opening night. Patricia Nicol (2005), reviewing the second Edinburgh run of the play, was one of many reviewers to consider *Mobile Thriller* an original play rather than an adaptation of an Italian text, 'An Italian import, though the vehicle is German. A monologue by Renato Gabrielli ... '. This critic went on to underscore the originality of the writing: 'But it is the mobile phone conversation that makes it fast and gripping. Which of us hasn't been witness to a stranger's private thoughts through overhearing them talking on their mobile? Technology meant to liberate us has actually altered irrevocably our sense of privacy and private space ... ' (Nicol 2005). Reviewer Sam Wollaston (2004), instead, commented on his excitement and surprise when he found himself playing the unusual role the spectator is expected to perform in *Mobile Thriller*: 'Happily this wasn't reality but drama, though it was so involving it was easy to forget. I've never felt so much that I was in the play rather than watching it. The suited man was an actor, the silent driver the play's producer. And the dark blue car was our playhouse.'

In Edinburgh *Mobile Thriller* won a Herald Angel Award and received a string of four and five star reviews, which doubtless helped it continue its travels. In 2005 it embarked on an intercity tour taking in London (Battersea Arts Centre), Warwick, Birmingham, Cambridge and Manchester and later in the year, Toronto. In 2005 it had another successful run at the Edinburgh Fringe, together with a short tandem play, *Broken Road*.

The changes I have discussed for the Edinburgh production have happened repeatedly as Carrie Cracknell and James Erskine went on to customize *Mobile Thriller* to each specific city. Without these continual adjustments to make the play fit the city where it is being performed, I do not believe it would have held such audience appeal or enjoyed such continuing critical acclaim.

The work of translating and co-adapting proved a brilliant learning experience. I was able to work more intensely with a company – director, writer and actors – than I had ever done in the past. Still the most important thing I learnt was that the translator/co-adapter must have the last word in decisions concerning the linguistic side of the adaptation with the aim of maintaining overall coherency in the play text.

The various roles I have described are separate but interlocking, assigning the translator-co-adapter-cultural mediator with more power in the negotiations room in the theatrical marketplace than is usually the case. Which plays are staged, and the reasons why a company or an artistic director selects one play rather than another for the following theatre season, depend on a host of different factors. A dominant one is undoubtedly the author's profile within the target culture. If, as I did, one can organize promotional events in the target country focusing on the author and her or his work, the chances of a play being selected for performance increase.

On the downside, even if one is willing to play diverse roles, a decision requiring a great deal of time and energy, in a bid to promote foreign plays abroad, my experience has been that one's work as translator is nearly always overlooked. When the translation works, critics simply tend to ignore the translator's role in the long process from page to stage. In the case of the 2004 productions of *Mobile Thriller* only one out of about twenty reviews mentioned my name.

With interest from producers and directors in Montreal, Tokyo, and Boston *Mobile Thriller* will hopefully soon be on the road again, in a new adaptation to suit these two very different cultural locations.

Conclusion

This article has explored how context of practice can determine the role that the theatre translator plays in the transfer of plays from one cultural and theatrical tradition to another. By following Rose's own experience in the translation, adaptation and British production of Renato Gabrielli's play, it has also suggested that translators can actively operate within those contexts of practice to affect the success of their translations. Rose acted not only as a cultural mediator between Gabrielli's play and its Scottish audience, but also as a cultural promoter, creating a market for the source play by raising funds for events that created a cultural environment amenable to Gabrielli's work and by building links with practitioners.

The articles has also shown that the translator is not only the initiator of the creative process but by becoming part of a creative team provides access to the source culture and the source text throughout the process of adaptation and staging, offering opportunities of cross-cultural exchange every step of the way. Finally, we hope that our aim of combining practitioner's account and critical reflection offers a convincing argument in favour of a study of theatre translation that engages with practice and takes into account the specifics of its complex dynamics of location, temporal situation and participants.

Notes

1. Italian academic practice requires a specification of the degree of the contribution of the two authors, which is as follows: Marinetti (30%) Rose (70%), with Marinetti having written the initial section and Rose the latter section.
2. For further information on the selection policy of the National Theatre see Bradley in this volume, and for more comment on the difficulty of promoting foreign theatre, see the interview with Christopher Hampton, also in this volume.
3. For details concerning Renato Gabrielli's stage work, see: http://delteatro.it/dizionario_dello_spettacolo_del_900/g/gabrielli.php
4. Renato Gabrielli's short story, *Qualcosa trilla*, was published in the *Maratona di Milano* collection in 2001. The author reworked this story into a play and kindly gave me the unpublished manuscript from which the quotations in the present chapter derive (Rose 2004).
5. Massimo's mobile phone conversations are central to the play, hence 'mobile' in the English title, while I decided to translate the Italian 'trilla' (trillare means 'to ring') as 'thriller' given the situation of mystery and suspense pervading the lead character's existence and death.
6. Translation one (T1) refers to the published play text, Translation two (T2) refers to the text which was performed at the Edinburgh Fringe in August 2004 and which evolved during discussions with the rest of the creative team.
7. See Cracknell, 2004. The show ran from 15 to 17 April, 2004.
8. James Erskine and Carrie Cracknell founded Hush productions in 2003. See: www.hushproductions.org. Carrie is currently artistic director at the Gate Theatre, London, together with Natalie Abrahami.

References

Aaltonen, S. (2000) *Time-sharing on Stage: Drama Translation in Theatre and Society* (Clevedon: Multilingual Matters)

Bassnett, S. (1998) 'Still trapped in the labyrinth: further reflections on translation and theatre', in S. Bassnett and A. Lefevere (eds.) *Constructing Cultures: Essays on Literary Translation* (Clevedon: Multilingual Matters), pp.93–109

Brown, M. (2003) 'Three Cheers for the New Alliance', *The Scotsman*, 14 May, p.11

Cracknell, C. (January–April 2004) *Tron Programme* (Glasgow: Tron Theatre)

Gabrielli, R. *Qualcosa trilla* (unpublished manuscript)

Mansfield, S. (16 April 2004) 'Death and the City' *The Scotsman* (online) Available at: http://www.highbeam.com/doc/1P2-13045337.html [Accessed on 20 February, 2008]

Puppa, P. (2006) 'The contemporary scene', in J. Farrell and P. Puppa (eds.) *A History of Italian Theatre* (Cambridge: Cambridge University Press)

Sipario: Mensile dello spettacolo. 718–719, 2009

Nicol, P. (15 August, 2005) 'Edinburgh theatre: raise your hand if you enjoy the Fringe', *The Sunday Times* (online) Available at: http://entertainment.timesonline.co.uk/tol/arts_and_entertainment/article468325.ece [Accessed 20 February, 2008]

Wollaston S. (18 August, 2004) 'The Fast Show', *The Guardian* (online) (Available at: http://www.guardian.co.uk/culture/2004/aug/16/edinburgh04.edinburghfestival9 [Accessed 20 February, 2008]

Rose, M. 2004 (trans.) *Death and the City: Mobile Thriller, Vocation* and *The Number Ninety's Child* by Renato Gabrielli, in *Plays International* (19:7), pp.43–4

9

Cow-boy poétré: a Bilingual Performance for a Unilingual Audience

Louise Ladouceur and Nicole Nolette

The two official languages of Canada have been historically confined to two linguistic territories, often referred to as the 'two solitudes',[1] with Francophones in the East, mostly in Quebec, and Anglophones in the centre and western parts of the country. According to statistics compiled by Canada's national statistical agency in 2006,[2] unilingual Anglophones make up 67.61 per cent of the population, while another 13.24 per cent consists of unilingual Francophones, residing for the most part in Quebec. The remaining 17.45 per cent of the population is bilingual, mostly Francophone, and is dispersed throughout the country, with approximately 2.17 per cent of them in the four provinces of Western Canada: Manitoba, Saskatchewan, Alberta and British Columbia. In these relatively small communities, far from the French centres in the East, bilingualism is a necessity of daily life for Francophones inasmuch as public exchange takes place exclusively in English, while French is confined to the private sphere, or to activities of the very few francophone cultural and educational organizations.

In this context, where French is a minority language constantly exposed to the predominant English of daily social interaction, Francophones have developed a specific vernacular that includes a French-English code-switching that alternates easily from one language to the other. This is a distinct language representative of the effects of asymmetrical bilingualism on the francophone minority communities of Western Canada.[3] For these francophone communities, eager to affirm the language in which their identity is rooted, theatre is seen as an act of cultural resistance, allowing the French language to resonate in the public sphere, thus affirming and promoting its existence.[4] The language promoted on the stage has often been an idealized, purified French, unadulterated by the interactions with English that are typical of the daily realities of life in a context where it is the dominant language. However, the past few years have seen the development of a francophone drama repertoire in Western Canada that has dared to display its own vernacular French-English code-switching onstage.

Previously a subject of shame, inasmuch as it was perceived as an inability to totally appropriate French, the bilingualism of French-Canadian communities is taking on another stature within the context of a globalization that bestows upon it an undeniable surplus value. No longer the symbol of an inability to exist as a real Francophone, it has become an indication of the ability to function in both French and English, a proficiency that entails indisputable economic and cultural advantages. Not only does it provide access to a global marketplace that has adopted English as a lingua franca, it reflects Canada's official bilingualism, a policy that enables Canada to differentiate itself from the United States by contrasting the Canadian multicultural mosaic to the *melting pot* of its powerful neighbour to the south. This bilingualism also acts as a conduit to international cultural networks that are constructed around either of the two languages, whether within the Commonwealth or the Francophonie. Institutionally, this bilingualism constitutes a 'distinctive feature' according to the sociological model of Pierre Bourdieu (1979, p.191), for whom 'l'identité sociale se définit et s'affirme dans la différence' ('difference is a key factor in defining and affirming social identity'). The bilingual aesthetic of the small, French-Canadian drama repertoires thus allows them to distinguish themselves from the decidedly-unilingual Quebec repertoire that occupies the centre of the francophone theatrical institution in Canada. Just as 'joual', the vernacular French language of Quebec, allowed Quebec playwrights to distance themselves from the normative French of France in order to develop a repertoire of their own, a strategy often used by minority literatures written in a prestigious literary language such as French or English (Casanova, 1999, p.73), it is by distancing themselves from the dominant norm to be found in Quebec that French-Canadian drama repertoires are able to highlight their specificity.

The bilingual aesthetic of plays that spring from the French-Canadian minorities also embodies sizeable disadvantages. Detracting from the idea that a culture defines itself by the awareness of its unity, a unity consolidated by language, multilingual works are destined to belong to no single literary system legitimately constructed around a single language. The situation is even more complicated in theatre where this heterolinguism, a term coined by Rainier Grutman (2006, p.18) to qualify the literary manifestations of multilingualism, acts as an impediment to the circulation of bilingual plays for audiences lacking the necessary linguistic competence to understand all of the languages involved. As underlined by Susan Bassnett-McGuire (1985, p.87), 'a theatre text exists in a dialectical relationship with the performance of that text. The two texts – written and performed – are coexistent and inseparable.' Subject, as it is, to performance imperatives, theatrical communication requires a message to be immediately understandable within the space and time of the performance, without recourse to dictionaries or footnotes (Ladouceur, 2005, pp.55–6). Because of the immediacy of the theatrical communication, '[w]henever a language or dialect appears in a production that is not likely to be understood by a majority of the audience, it requires some sort of performance adjustment if full or general communication is desired' (Carlson, 2006, p.181). Inasmuch as only a small percentage of the Canadian population is bilingual, the heterolingual dialogue of French-Canadian plays predestines them to marginality while circulating on the periphery of the major unilingual French or English theatre centres in Canada and abroad. It is a vicious circle: the very element that contributes to a distinct cultural product in French acts simultaneously as a source of discrimination in the marketplace of French cultural exchange. To escape this discriminatory effect towards the bilingualism they incorporate, these plays are faced with the need to be translated. This is a condition *sine qua non* to access a larger French theatre scene that is mostly unilingual and to penetrate the markets governed by the dominant francophone institutions. But how does one translate a bilingual text to render it accessible to an audience ill-equipped to understand both languages without erasing the linguistic duality that is at the very heart of the French identity it wants to express?

Until recently, the French-Canadian plays that have been translated into English are those that exhibit a minimal bilingualism that can be easily reproduced in English. Such is the case for Franco-Ontarian Jean Marc Dalpé's plays *Le chien*, *Lucky Lady* and *Trick or Treat*. Written mostly in French, the texts contain a few terms or phrases in English that provide non-essential information and, therefore, carry no diegetic[5] value. The English translation basically reproduces the linguistic profile of the original by adding, where the context will allow for it, a rare French term or expression devoid of any information necessary to the understanding of the play. If Dalpé's plays readily lend themselves to this type of translative approach, it is because they contain an accessory heterolinguism in which the passages delivered in the other language have little or no pertinent significance. However, this model is not applicable to texts with major heterolingual content, such as the bilingual plays recently produced in the remote francophone communities of Western Canada.

The itinerary of *Cow-boy poétré*

Created in Edmonton, Alberta, in April 2005, *Cow-boy poétré* exhibits a bilingualism in each and every step of its creation and production. The title is an eloquent symbol of the back-and-forth exchange between the two languages employed in the text. It is comprised of two English terms that have been Frenchified: the first, *Cow-boy*, employs the hyphen specific to the French spelling of the word, while the second is a phonetic translation into French of the word *poetry*. *Cow-boy poétré* refers to the custom, as old as cowboys themselves, of sitting around a fire after a hard day in the saddle in order to tell stories or recite poetry. This tradition has spawned a genre referred to as 'cowboy poetry' that flourishes even today.

The genesis of the play reveals a great deal of a context specific to the Canadian prairies. The idea originated with Daniel Cournoyer, the director of L'UniThéâtre, the only professional French theatre in Alberta. He wanted to create a play that dealt with the rodeo because he viewed the rodeo as an important element of the rural Albertan psyche and something he wished to portray in a play that would fulfill his theatre's mandate: 'to validate, via a repertoire of original works, a Francophonie that isn't Quebec or France, that lives and breathes the West' (Nicholls, 2005, p.C2). After fruitless attempts at finding a francophone author, Cournoyer called upon an Albertan anglophone playwright whose work was rooted in Western Canada and who was well acquainted with the rodeo world. Kenneth Brown accepted the commission and produced a first version in English that was followed by a first reading in English. The revised text was then given to Laurier Gareau, a francophone author and translator from the neighbouring province of Saskatchewan, who was entrusted with fashioning a partial translation, inasmuch as approximately one quarter of the play was kept in English.[6] This portion consists mainly of the presentation of the rodeo events and participants by the announcer and of various musical performances, all of which are delivered in English in an attempt to maintain authenticity, rodeo being a public activity that takes place in English in Western Canada even when the participants are Francophones. The other four characters are bilingual Francophones speaking a Western-Canadian 'Franglish', a French vernacular containing numerous English words and phrases and, in this case, all the terms related to ranching and rodeo events, as can be seen in this excerpt (Brown, 2005, p.3):

> Luke: J'aurais pu être le meilleur ostie de *bull rider* n'importe où au monde. Vous me voyez astheure pis vous vous dites « ce p'tit pisseux de *gimp* y pourrait pas *rider* une clôture ». L'monde y sont comme ça (. . .) L'monde pense que les *bull riders* sont des gros hommes. Y nous prenne pour des osties de *steer wrestlers*. J'peux vous dire qu'y en pas d'*bull riders* plus haut que cinq pieds huit . . . pis ça c'est sans bottes. Mais c'est qui les osties d'stars du rodeo? Ben j'peux vous dire que c'est pas ces gros bâtards là qui s'coltoyent avec des pauvres imbéciles d'animals sans gosses. Pis c'est surtout pas les p'tits fifis d'*bronc riders*. (my italics)

The back and forth between the two languages continued during rehearsals as all major revisions to the text were done in English, then translated into French when needed. Thus, a complex activity of translation and interweaving of languages was shaping not only the text but the whole process of creation from which it sprang.

Of the five characters in the play, four are bilingual Francophones: the country western singer Chantal and the three cowboys, Blanchette, Diamond and Luke. The fifth character is the rodeo announcer who speaks only in English. In the written text, the level of language spoken by the four bilingual francophone characters is similar, except for the curses that are liberally sprinkled throughout Luke's dialogue, curses that are characteristic of a Quebec slang but almost non-existent in the lines delivered by Chantal, Blanchette and Diamond, who are Albertan in origin and are thus more deferential in the face of religion. The oral text, however, clearly confirmed the origins of the francophone characters through their accents. In the production of 2005, Luke displayed a decidedly Quebecois accent, while a 'western' pronunciation could be detected in the speech of Blanchette and Diamond. As for Chantal, her accent took on an almost mythical proportion since the character was played by Crystal Plamondon, a veritable legend of country western music, who divides her time between Alberta and Louisiana. Incorporating sonorities from the Southern United States and from Western Canada, Chantal's speech alone traces out a map of the North American francophone diaspora.

This multitude of accents constituted a complex 'heterophonia' that underlined the characters' origins and was a powerful dramatic component in a plot where francophone cowboys compete for the favours of a beautiful, country western singer, who will end up leaving her Albertan husband for a lover from Quebec. In the Albertan context, this amorous rivalry is laden with strong, ideological and political connotations inasmuch as it reproduces the power struggle that francophone communities are caught up in, wherein Quebec is dominant and the other francophone regions of Canada are, for the most part, neglected. For instance, in much official discourse and many documents, Francophones in Canada fall into two categories: Quebecers and the Francophones outside Quebec, implying that the second group lacks a specific identity other than not being a Quebecer.

The play's direction and set design were also indicative of the linguistic dynamic that exists in the francophone context of Western Canada. The playing area was divided into two zones: the first zone, on ground level, was shaped as a rodeo infield, covered with earth and dotted with bales of hay, while a second was comprised of a platform upon which stood the announcer and the musicians whom Chantal would join when she sang. Representative of the public sphere, the platform was reserved either for the announcer, who spoke only in English when introducing the various rodeo events and providing a running commentary on them, or for the musical numbers with Chantal and the musicians. All the private scenes took place in French in the first zone which served alternately as a pasture, the interior of a house, a hotel room and a hospital corridor. This zone became public space, and dialogue was delivered in English, only when it was used for the rodeo action. For instance, with the entrance of the mechanical bull upon which the cowboys will prove their worth, a spotlight isolated the bull, thus linking it to the public space on the platform where all the dialogue occurred in English. It also happened when the rodeo clown came out to entertain the audience while the bull was removed from the stage. Therefore, these two playing areas spatially illustrated the divide between English, the common language of public life that unfolded on the platform, and French, the language devoted to private exchanges that took place in the playing area at ground level.

This production was subsequently presented in Ottawa on 9–10 November 2005 as part of the *Festival Zone Théâtrale*, a festival destined 'to showcase professional theatrical productions originating in Canada's Francophone minority communities (for instance outside the Province of Quebec) and in various regions of Quebec' (National Arts Center website). The play also toured the province of Alberta and journeyed to New Brunswick, where it was presented at the Théâtre national d'Acadie and the Théâtre de l'Escaouette in September 2006. Composed mainly of bilingual francophone spectators living in minority communities in Central and Eastern Canada, these audiences could easily understand the heterolingual dialogue of the play. The main obstacle they faced in relating to the production was its subject matter. Rodeo being an activity specific to Western Canada and alien to them, spectators commented upon the difficulty they had in understanding the specifics of each rodeo event and appreciating what was at stake in dialogue focusing on various aspects of the rodeo life. This is a concern that could be addressed through various translation strategies destined for audiences unfamiliar with the play's subject matter, strategies such as those proposed in this essay.

Bilingual performance for a francophone audience

As the language of *Cow-boy poétré* reflects the diglossia inherent to francophone communities in Western Canada, it is not easily understood by an audience that is unaware of this reality, as is the case for most francophone spectators in Quebec. In order to gain access to the stages of Quebec, where the francophone theatre institution enjoys the most prestige and highest degree of legitimacy, the play must necessarily undergo another process of translation. However, if the translation of the bilingual text must render it accessible to an audience lacking the prerequisite linguistic knowledge of both languages, it should also maintain the linguistic duality that is characteristic of the community it represents. Therefore, it requires translation strategies able to maintain the heteroglossic value of the original production. Such strategies can involve 'on-stage translation' (Carlson, 2006, p.181) devices that 'are consciously brought into the theatrical frame' (Carlson, 2006, p.181). The use of surtitles[7] for instance, where written simultaneous translation is provided on stage, is already prevalent in international theatre festivals where it allows entries from different countries to accommodate a multilingual audience. As is the case with subtitles in the film industry, surtitles offer audiences the possibility to enjoy a theatre production in its original form and to understand foreign works that would otherwise be unintelligible. It is a target-specific mode of translation, able to provide performance adjustments specific to the audience to which the production is destined.

The theatrical act in itself can also be used as the matrix for on-stage translation devices that go beyond the textual to include performance in action. Such a performance-based process can be observed in the multilingual productions that Canadian artist Robert Lepage[8] presents on various international stages. For instance, in *The Seven Streams of the River Ota*, Lepage uses various on-stage translation strategies for a dialogue that mixes French, English, German and Japanese. When the language delivers non-essential information, no translation is provided, but when it is necessary to understand the dialogue, a character is transformed into a translator onstage and simultaneously translates into English a phone conversation taking place in Japanese and French. Elsewhere, English versions of German or French dialogue are displayed in surtitles on a screen that is either an integral part of the stage or placed above it. These devices are introduced into the original production in order to adapt it for a specific audience. In contrast to the published text, which is immutable in its printed form, the theatricality of the work on stage allows it to be modified to suit changing audiences. This is a fundamental characteristic of dramatic works that translation can make use of to explore translative strategies no longer attuned to the printed page but rather to the performance on stage, shifting from literary to theatrical concerns.

Thus, a production of *Cow-boy poétré*, destined for a unilingual francophone audience in Montreal, could be adapted according to the following model.[9] For an audience whose contact with English is habitually indirect, French being the sole official language of Quebec, as well as the dominant language in the public sphere, the original French dialogue could be kept as is, while the English dialogue would be subject to different on stage-translation processes, including surtitles and performance-based devices.

Surtitles

To begin with, for an audience unfamiliar with the rodeo, it would be necessary to provide surtitles or an illustration that would detail the names and nature of the rodeo events in English and in French. This would allow for a much better understanding of what is at stake in the play for spectators unfamiliar with the specific realities of rodeo. A bilingual glossary and description of rodeo terms could as well be included in the program, along with the rules pertaining to each event. For example, Bull Riding/*Monte du taureau* is the most dangerous of the rodeo events. The cowboy must stay on the bull for eight seconds and, to avoid disqualification, he must ride the bull using only one hand, while ensuring his free hand neither comes into contact with himself or the bull. This information would certainly enhance one's appreciation of the scene with the mechanical bull and of the play as a whole.

Performance-based translation devices

Several excerpts from the play could be the object of various translative strategies intended for a unilingual Francophone audience. In the original version of the following excerpt, the announcer introduces the rodeo events in English. For the translated version, not only could French surtitles be used, performance-based translation devices could as well be inserted into the matrix of the original production in order to modify the way it is performed, thus creating another version of the play accessible to a specific audience. In this manner, the role of Chantal could be enlarged to include dialogue containing a French equivalent of the message delivered by the announcer, although in a more succinct and personal manner, retaining only what is essential for the understanding of the action. She would thus become the francophone transmitter of the announcer's English dialogue for that portion of the text, as shown in this excerpt taken from the 2005 unpublished manuscript of the play available at l'UniThéâtre.[10] The proposed translation might appear as such, with the added text appearing in bold type:

Annonceur: Folks, one of our favorite rodeo cowboys, Jack Blanchette, out of Longview, Alberta. Jack has had some bad luck on the rodeo circuit this year. Finished out of the money last week at the Patricia Stampede Days, and didn't quite make the cut for the big rodeo in Calgary. But it ain't for lack of trying, and he's one tough hombre. Jack is riding Buckaroo today, and watch out cause he's one hell of a horse.

Chantal: **Et maintenant, un de nos meilleurs cow-boys, Jack Blanchette de Longview, Alberta. Jack a pas été chanceux c't' année dans l'circuit pis y s'est pas rendu au Stampede de Calgary. Mais c'est pas parcqu'y a pas essayé. Jack y'a la couenne dure pis aujourd'hui y va monter Buckaroo, un cheval qui a pas l'habitude de se laisser faire. Vas-y Blanchette, montres-y de quoi t'es capable!**

Annonceur: You ready boys? Alright, give her a ride, boy. (*The starting bell is heard.*) There he goes. Hang on there, cowboy, that's one hell of a rough rider. That's why they call them Roughriders, girls. Hang on . . . (*Blanchette is thrown from his horse. We hear the sound of a hoof hitting bone. With a sudden movement his head is jerked backwards as though he has been hit by the horse's hoof.*

	The crowd groans at the sound of the bell.) Oh boy, folks, that's a hell of a kick to the head. Our Cowboys are try'n to get Buckaroo to pay some attention to them. Way to go guys. Is he okay, boys? (*Staggering, Blanchette gets up and heads towards the fence.*) Folks, anybody tell you that rodeo cowboys ain't tough, you tell me how many men could take a kick to the head like this fella and stand up fifteen seconds later. Give him a big hand, folks, it's the only pay he's gonna get today.
Chantal:	C'tait toute une ride pour Jack, mais y est capable d'en prendre. Après le kick qu'y a eu su à tête, y s'est relevé ça a pas pris 15 secondes. On l'applaudit ben fort, y faut être tough pour vrai pour encaisser un coup de même, pis c'est tout ce que Jack va encaisser icitte aujourd'hui.

The translation process is relatively straightforward in this excerpt as it is a monologue in which there are no direct exchanges between characters. It becomes much more complicated when a scene involves more than one character. Not only must Chantal be included in the scene, but some lines must be rewritten and others redistributed in order to render the message understandable in French, all the while retaining a natural flow in the dialogue with an eye on the clock so as to not unduly extend the length of the scene. For example, when Luke interacts directly with the announcer, he can reply either in French or in English, while repeating part of the announcer's questions in French in his answers. Other utterances can also be moved, removed or added to allow the audience to immediately understand what the characters are saying.

For instance, there is a scene in which Luke, nicknamed 'the Clown Prince of Rodeo' for his ability to entertain the audience between events, performs a routine pertaining to the political opinions in Western Canada. He mentions Pierre Elliott Trudeau, a Canadian prime minister known for implanting bilingualism and multiculturalism policies in Canada in the seventies and early eighties. Trudeau is not a very popular figure in Alberta because of his National Energy Program, which was seen as depriving Western Canadians of the full economic benefits of 'their' oil and gas. This same prime minister is also controversial in Quebec, but for different reasons, notably the imposition of the *War Measures Act* and the repatriation of the Constitution without Quebec's consent. A performance-based device could take advantage of this shared perception in producing a translation destined to a Quebec audience. The following excerpt contains the dialogue, originally taking place in English between Luke and the announcer in the 2005 unpublished manuscript version of the play. It has been synthesized and redistributed to include additional lines in French delivered by Chantal. The added lines appear in bold type in the text:

> (*Luke enters with an effigy made of straw and a broom.*)
> Annonceur: Say, Luke, how do you tell when a Liberal calf is born?
> Luke: **Comment c'qu'on peut savoir que c't'un veau libéral qui vient d'naître? Parce que ça sort du mauvais trou!** Comes out the wrong hole, Mr. Stockman.
> Chantal: **A puait pas mal celle-là!**
> Annonceur: Guess that's where the smell comes from! Who's your friend, Luke?
> Luke: **Ben icitte j'ai mon ami M. Trudeau.** I got my pal Mr. Trudeau right here, Mr. Stockman!
> Annonceur: How's that? You mean the politician? What's the cane for? Is somebody going to be limping outta here?
> Chantal: **Ben, dis-nous donc à quoi à va servir, ta canne.**
> Luke: Oh no, Mr. Stockman, I had something else in mind. You know what the Liberals been doing to the West for the last ten years?
> Annonceur: Yup. But it's not polite to say in mixed company.
> Chantal: **C'que Trudeau fait à l'Ouest depuis dix ans, ça ressemblerait-tu à c'qu'il fait au Québec?**
> Luke: What, you mean THIS? (*He gooses the effigy with the broom. The crowd roars.*)
> Annonceur: That's some medical operation. What do you call that?
> Luke: I call it parliamentary debate, Western style!
> Chantal: **Ca c'est ce qu'on appelle le débat parlementaire, à la Western!**
> Annonceur: Ain't he something, folks? That's Luuuuuke Kaaaaannne, the Clown Prince of the Western Canadian Rodeo!

The steps the text has gone through in the translation process are summarized in this Table 9.1, showing the three different versions that have been developed for different audiences. It went from the original version

Table 9.1: Language use in various versions of the play *Cow-boy poétré*

English text	Bilingual text for bilingual audience	Bilingual text for Francophone audience
Written text: – English	*Written text:* – Franglish (franglais) – English	*Written text:* – Franglish (franglais) – English
	Oral text: – Franglish, Alberta accent – Franglish, Quebec accent (joual) (more swearing, less code-switching) – English, Western/Central Canada	*Oral text:* – Franglish, Alberta accent – Franglish, Quebec accent (joual) – English, Western/Central Canada ↓ On-stage translation: <u>Surtitles</u> – French surtitles of rodeo events <u>Performance-based translation devices</u> – presentation of rodeo events Chantal as French announcer (addition of text in French) – rodeo clown (redistribution of already existing text and addition of text in French)

written in English only and not intended to be performed, to a bilingual translation written and performed in both languages for bilingual francophone audiences and, finally, to a bilingual translation destined for a unilingual francophone audience. In the third column, we can observe that the written text is different than the oral text, in which the accents serve to distinguish two types of French, a vernacular Albertan French or 'Franglish' and a French vernacular spoken in Quebec, referred to as 'joual'. Only the English dialogue is translated, but it is a partial translation done with the help of on-stage translation including surtitles and 'performance-based translation devices' developed for a specific performance to meet the requirements of a specific target audience and integrated into the previously bilingual version. For a production in Montreal, for instance, surtitles could be useful to deliver information on the rodeo events integral to the plot and essential to an understanding of the action. Performance-based translation devices could include the insertion of Chantal's commentary in French in the English dialogue of the announcer, which would involve additional text in French and a redistribution between Chantal and Luke of already existing text.

Conclusion

In Western Canada, where recent dramatic works display an accented bilingualism, representing an inescapable reality for Francophones living in a minority situation, new forms of translation must be used to preserve the linguistic duality of the bilingual work while accommodating an audience with different language skills. With the use of surtitles and performance-based translation devices, translation can benefit from the performative resources specific to theatre in order to adapt the original production for the intended audience. The selection of these performance-based translation devices is made to accommodate a specific target audience and their insertion into the original play constitutes its translated version. Thus, the translation is anchored in the original play which serves as a matrix integrating all the variations necessary to make it accessible to a specific audience. In this instance, this allows the work to access the larger francophone theatrical centres of Canada and abroad, without losing the linguistic specificities of the community from which it emerges, and which sets it apart from other Franco-Canadian communities. This mode of translation specific to theatre is able to explore all the resources of the theatrical act to enhance the circulation of plays on the stages of the world while maintaining their originality and remaining faithful to the context of their creation. In this case, it is on the Prairies of Western Canada, where small French communities struggle to preserve a language in which their identity is rooted, while living their public life in English and even embracing the lingo of the cowboys and cowgirls who inhabit Canada's Far West.

Notes

1. From Hugh MacLennan's novel *Two Solitudes*, published in Toronto by Collins in 1945. The title is a metaphor that has become the emblem of the relation between francophone and anglophone cultures in Canada. For more details, see Ladouceur 2009.
2. See Statistics Canada web site: http://www40.statcan.gc.ca/l01/cst01/demo 15-eng.htm

3. For a study of the various degrees of bilingualism exhibited in French-Canadian drama and their English translation, see Ladouceur 2006.
4. For more details, see Ladouceur 2005.
5. From the Greek word *diegesis* meaning narration. It qualifies various narrative devices within structuralist discourse analysis. See Genette, 1973, 1980.
6. For more details on the genesis of the play, see Ladouceur 2006.
7. Marvin Carlson uses the term 'supertitles' for 'surtitles'. In this essay, we have opted for the more commonly-used term 'surtitles'.
8. For a discussion of the translation strategies displayed in Lepage's productions, see Ladouceur 2006, Koustas 2003, 2004.
9. The translation strategies put forward in this article serve only as examples and have yet to be integrated into an actual production.
10. The play has yet to be published. A manuscript version is available at l'UniThéâtre in Edmonton.

References

Bassnett-McGuire, S. (1985) 'Ways Through the Labyrinth, Strategies and Methods for Translating Theatre Texts', in T. Hermans (ed.) *The Manipulation of Literature, Studies in Literary Translation* (London: Croom Helm), 87–102

Bourdieu, P. (1979) *La distinction: critique sociale du jugement* (Paris: Minuit)

Brown, K. (2005) *Cow-boy poétré*. Partial French translation by Laurier Gareau, produced at l'UniThéâtre in Edmonton, 7–17 April 2005, unpublished manuscript available at l'UniThéâtre (Edmonton)

Carlson, M. (2006) *Speaking in Tongues: Language at Play in the Theatre* (Ann Arbor: University of Michigan Press)

Casanova, P. (1999) *La république mondiale des lettres* (Paris: Seuil)

Dalpé, J. M. (1987) *Le chien* (Sudbury: Prise de parole)

Dalpé, J. M. (1995) *Lucky lady* (Montréal: Boréal)

Dalpé J. M. (1999) *Trick or treat*, in *Il n'y a que l'amour* (Sudbury: Prise de parole)

Genette, G. (1973) *Figures III* (Paris: Seuil)

Genette, G. (1980) *Narrative Discourse,* J. Lewin (trans.) (Ithaca and London: Cornell University Press)

Grutman, R. (2006) 'Refraction and recognition: Literary multilingualism in translation', *Target, International Journal on Translation Studies*, 18.1, 17–47

Koustas, J. (2003) '*Zulu Time*: Theatre beyond Translation', *Theatre Research in Canada*, 24.1–2, 11–21

Koustas, J. (2004) 'Robert Lepage's Language/Dragons' Trilogy', *International Journal of Canadian Studies/Revue internationale d'études canadiennes*, 30, 35–50

Ladouceur, L. (2005) *Making the Scene: la traduction du théâtre d'une langue officielle à l'autre au Canada* (Quebec: Nota bene)

Ladouceur, L. (2006) 'Write to Speak: accents et alternances de codes dans les textes dramatiques écrits et traduits au Canada', *Target, International Journal of Translation Studies*, 18.1, 49–68

Ladouceur, L. (2009) 'A Firm Balance: Questions d'équilibre et rapport de force dans les représentations des littératures francophones et anglophones du Canada', *From a Speaking Place: Writings from the First Fifty Years of 'Canadian Literature'*, 128–45

Lepage, R. and Ex Machina (1996) *The Seven Streams of the River Ota* (London: Methuen Drama)
National Arts Center, http://www.nac-cna.ca/en/news/viewnews.cfm?ID=981&cat=catFT, date accessed 15 August 2009
Nicholls, L. (2005) 'Allo, mon cowboy', *The Edmonton Journal*, 6 April, C1–C2
Statistics Canada, http://www40.statcan.gc.ca/l01/cst01/demo15-eng.htm (Accessed 15 August, 2009)

Part III
In Conversation with Practitioners

10
Interview with Christopher Hampton

Roger Baines and Manuela Perteghella

Christopher Hampton is a playwright, screenwriter, director, producer and translator. His stage translation and adaptation work has won numerous awards and includes Chekhov's Uncle Vanya *(1971),*[1] Three Sisters *(2005), and* The Seagull *(2007), von Horváth's* Tales from The Vienna Woods *(1977),* Don Juan Comes Back From The War *(1978),* Faith, Hope and Charity *(1989), and* Judgement Day (2009), *Ibsen's* A Doll's House *(1972),* Hedda Gabler *(1972),* The Wild Duck *(1980) and* Ghosts *(1983), Laclos'* Les Liasions Dangereuses *(1985), Márai's* Embers *(2006), Molière's* Don Juan *(1973) and* Tartuffe *(1984), and Reza's* Art *(1996),* Life x 3 *(2001),* Conversations After A Burial *(2007), and* God of Carnage *(2008). Interview by Roger Baines and Manuela Perteghella (autumn 2008).*

RB: **What was the first piece that you adapted for the stage?**

CH: The first thing I did as a sort of a translator was in 1967: the Royal Court did a play called *Marya* by Isaak Babel, translated by Michael Glenny and Harold Shukman. It was directed by Robert Kidd who had directed my play (*When Did You Last See My Mother?* [1966]) and they were into rehearsal and they, the actors and Robert, felt that some more work needed to be done on the translation so they called me. I was still a student at Oxford. It was very interesting actually to go and sit in rehearsals. I worked with Michael Glenny and not so much with Harold because he was away. There was a slight awkwardness at the beginning, because Michael was a very distinguished Russian translator and he was some sort of don at Oxford and I rang him up and said I've been asked to do this and he was very nice; and then I said it's handy because I live in Oxford and he said what do you do and I was obliged to tell him that I was an undergraduate. However, when I got to know him he was very helpful and entered into the spirit of the whole thing and he revealed that he and Harold Shukman had literally parcelled up the play in scenes and had translated alternate scenes, so it wasn't surprising that there was no coherent line through the play for the actors. Anyway, I was more or less coordinating with Michael and scribbling on the backs of envelopes as the rehearsals went on and that was my first real taste of translating. It was the case that the Royal Court was pioneering this idea of writers, playwrights doing adaptations of the classics. And they particularly embarked on it vis-a-vis Chekhov. They had asked Ann Jellicoe to do *The Seagull*, Edward Bond to do *Three Sisters*, and they asked me, when I went from Oxford to work at the Royal Court as resident dramatist, to do *Uncle Vanya*. Anthony Page directed it. I had studied French and German so I was technically a linguist, but I had only done one year of Russian and I didn't even take the O-level at school, so I sort of knew the alphabet, but very little else. So Anthony Page found a woman called Nina Froud, who was the editor of *The Penguin Russian Cookbook* and lived somewhere in North London, and she did a literal translation, which I asked her to make as literal as possible, to observe as closely as possible the punctuation and the length of the sentences, which she did, and we did it an act at a time over several months. I would do, as it were, the first act, and we would go and spend an evening at Nina's and she would cook some fabulous meal and we would go through it pretty much line by line, because the intention was, and actually my intention always is when I translate, to try to get as close to the original as possible and to try to reproduce its effects as closely as possible, so that's what we did over several months; and then we went into rehearsal with Paul Scofield, Colin Blakely, Anna Calder-Marshall. We pretty well did what the Royal Court used to do, which is that not a word was changed. It's not the same today. But the philosophy of the Court then was that the author's text was holy writ and nobody could change anything. I remember once, I plucked up the courage to ask Paul Scofield why he had delivered a line in such an eccentric way and it turned out that he was observing the punctuation and I had left out a comma.

MP: **So, there was no input from actors?**

CH: No, not at this stage. The very next thing I did was *Hedda Gabler*, again from a language I don't know. Again, I got a literal translation from a woman in Golders Green called Hélène Grégoire and that was a very rushed job because Peter Gill was directing the play at Stratford,

Ontario, and he had commissioned another writer who had defaulted or something, so I literally had about five weeks to do it and so I did it as fast as I possibly could and pretty well was putting the final full stops on the way to the airport. There was a lot more input on that production. Irene Worth was playing Hedda Gabler and she had a lot of . . . it's been my experience that people who play Hedda Gabler tend to be quite strong-minded in one way or another. Anyway, when I arrived in Stratford at the beginning of rehearsal, she suggested that it might be interesting to explore some byways and to expand this or contract that and I absolutely refused, I was very puritanical about it, whereupon she became very frosty for a while, but in the end it was a very successful production and I think everybody felt that it had paid dividends sticking to the text. I believe we made a few cuts, as tends to happen with Ibsen, you can make a few cuts without anything material being damaged. Looking through the text with Hélène Grégoire, we started to see patterns of words, we started to pick out words that belong to certain characters or that recur. In *Hedda Gabler* a word that struck us . . . I think it's 'tore' which means to dare. This word is used as a counterpoint throughout, so that you realize as you work through the translation that it's one of the central themes of the play. Hedda is fundamentally dissatisfied with life, but what she's particularly dissatisfied about is that she doesn't dare do this, that or the other. In other words, the play is about her failure of courage, which she can only resolve by taking the daring decision to kill herself. Working on *Uncle Vanya* and *Hedda Gabler* I began to realize that you need to pay very close attention not only to what the lines mean but to the music of the lines and the specific use of various expressions. Chekhov is very fond of a phrase which means 'if you only knew', a melancholy expression that's used quite musically throughout *The Seagull*, which I have just done, but also in *Uncle Vanya*.

So, I was launched as a translator pretty much by chance. It was a great stroke of fortune for me because there was a period of time, roughly the 1970s and the 1980s, where, if you did a translation and there was a successful stage production of it, it became the translation that people used; so therefore, in America, certainly, whenever *A Doll's House* or *Hedda Gabler* were done, it tended to be in my translation, which was great. Now it's gone completely to the other extreme and theatres nearly always insist on a new translation, every single time. And a lot of playwrights are now making this quite a significant part of their working practice.

MP: Do directors require a different translation for their own production?

CH: Yes, and often Trevor Nunn or Peter Hall, for example, will do it themselves and work in the rehearsal room with the actors. I think people are now firmly wedded to the idea of renewing the franchise every time. Even with *The Seagull,* which was exceptionally successful in New York, I don't anticipate it becoming the standard translation, because there are so many versions lurking around particularly of the Chekhov plays, there's an enormous choice, and a lot of them are published. So everybody from David Mamet to Tom Stoppard to Michael Frayn, they've all done their share of Chekhov.

RB: That was your way in, but you do a lot of very different work. You write your own work, you write adaptations but you still do the translating. So even if you got into it by chance, could you say why, what is it that draws you to translating for the stage?

CH: I find it a very valuable exercise. And I enjoy it. It's a real contrast from what you might call original work because it's such a strict template, especially if you do it with my attitude or philosophy. It's a discipline, it's like going to the gym. It works that writing muscle. I practically always have a translation on the go, which I do, as, I suppose, a pianist would do exercises. You spend two, three months working on a masterpiece and then into rehearsal with it, you sit and observe how the watch is put together and hopefully, it helps your own work in some sort of way. It's a real privilege to be able to work on these plays and see how they're made. Writing plays is so difficult. So few people have actually managed to do what Chekhov or Ibsen did, that it's tremendously valuable to try to analyse it. And you find that writers are subject to stereotyping or misunderstanding. Chekhov is so often portrayed as lyrical, sad and melancholic. He certainly doesn't shy away from the tragic aspects of life, but the language is very crisp and lucid, not fudgily poetic. In the famous final speech of *Uncle Vanya,* Sonya speaks of 'a whole sky of diamonds', she doesn't say 'the stars glinting like diamonds'. It's a very blunt play all the way through. The first line of *Uncle Vanya* is the nanny saying 'tea dear' and it's usually translated as 'have a cup of tea, dear' or 'why don't you sit down and have a cup of tea with me?', where in the Russian it's just those two words. You need to pay attention to these signs.

RB: This is in some way saying that as a translator you are a cultural representative for these texts, and you should respect them

as much as possible and that to some extent Chekhov has been misrepresented.

CH: I think so, and I think Ibsen also, you know, 'the gloomy Norwegian'. A play like *Ghosts*, for example, which is thought of as the epitome of doom and gloom is actually quite amusing. It's full of legitimate places for the audience to laugh, if you let them. Famously Chekhov was always complaining that the comedy in his plays was overlooked or somehow suppressed by Stanislavski. You try to pay attention to what the author's intentions might have been. We did *Hedda Gabler* at the end of the 1980s at the National and I revised the translation quite extensively, because I'd found by then a new method of translating Ibsen which was to track down the original German translation. Now, Ibsen was in Germany for nearly twenty years, and a lot of those plays were premiered in Germany, and he supervised the translations very carefully. So if you sit with the Norwegian and the German and you speak German, you can do it without recourse to a literal translation, you can do it directly, although I always have a literal translation just in case, just really to have someone to talk to about the various problems that might crop up. As we said, there is a proliferation of versions of most plays around and for me the fatal thing would be to look at anyone else's, because then you start to think: that's rather good. So it's best not to look at them at all.

MP: **And do you find translating from the Norwegian or the Russian more challenging? Do you feel as though there is someone else who is mediating?**

CH: Yes, I find it more difficult and in fact the weird thing was that towards the end of my time at the Royal Court, I said to Bill Gatskill you keep asking me to translate from all these languages that I don't understand. And he said what would you like to translate? I wanted to translate *Don Juan* by Molière, because as far as I knew at the time, which was the early 1970s, it had never been done, so that was my first translation of a play I knew well and that I chose myself. Naturally, in quintessential Royal Court fashion, when it came in, they didn't want to do it, but eventually it got done on the radio and the Bristol Old Vic took it up and did it, but it was never picked up by the Royal Court. Anyway, I felt much more comfortable with a French play; Molière and Racine were my special subjects at Oxford.

MP: **You had direct access.**

CH: Yes, and I knew the subject so well. It doesn't particularly, in the end, make it any easier, because there are individual problems with every play. One of your questions had to do with whether it was necessary to feel any affinity with the author. I absolutely have to. I've translated five authors, not counting Isaak Babel and Feydeau: Chekhov, Ibsen, Ödön von Horváth, whom I love (I've just finished a new translation, *Judgment Day*, which the Almeida are doing in 2009), Molière and, of course, Yasmina Reza, who's actually the only one who's liable to turn up at rehearsals and quarrel with you.

RB: **You talk about translation but obviously you adapt as well. Would you say that it's a different process, taking something like Laclos'** ***Les Liaisons Dangereuses?***

CH: For one thing, there is hardly one line of dialogue in a four-hundred-and-fifty-page novel and furthermore, Merteuil and Valmont, the two main characters, never meet, so you really just take the plot, gratefully, and then write a play, trying to keep as close to whatever it is that attracted you in the original. I like that process. What I'm doing right at this moment is adapting a Horváth novel. Toward the end of his life, harried by the Nazis, he couldn't get his last plays performed, so he began to write novels. His greatest success was one called *Jugend ohne Gott* (*Youth Without God*). I'm adapting it for the Theater in der Josefstadt in Vienna.

RB: **How did you get the rhythm out of the Laclos, making decisions between very ornate and archaic language but also a modern idiom?**

CH: I went for an eighteenth-century syntax and a more modern language. I've been away in New York and I went to see *The Seagull* again, and at the end, one of the two women sitting in front of me said: 'It's a wonderful production but I don't know about the translation'. She said: 'I mean, "Get a grip", they wouldn't have said that', which is a fair enough point, but you have to make these choices. I had a similar controversy with *Uncle Vanya* when Vanya arrives with flowers for Elena and finds her kissing the doctor and says something in Russian which I translated as 'Not to worry' – which is sort of what the Russian means, 'Don't disturb yourself' – so I said 'Not to worry'. Lots of critics picked up on it. As far as we could establish, 'Not to worry' was an expression that came in during the First World War. Anyway, I was so unrepentant that I put 'Not to worry' in *The Seagull* and nobody said a word about that. I think there's a Trevor Griffiths translation of a Chekhov play which contains the line: 'Up yours, butterballs' which may be a step too far. I think you have to be judicious. The most conspicuous example of this was when I did *The Wild Duck* for the National, which Christopher Morahan directed in 1979. I worked with Karen Bamborough, whom I knew already in her capacity as a Channel 4 producer, her mother was Norwegian. She said to me, at the very end of the play, the doctor has 'a plague on both your houses' kind of line and in nineteenth-century Norwegian, there isn't really any swearing, but what the doctor says at the end of the play is the worst thing you could possibly say, the strongest thing you could possibly say. So I translated it as 'go and fuck yourself'. In the previews it became clear that the audience was very thrown by this and we had a meeting, Peter Hall, Christopher Morahan who was directing it, and Basil Henson who was playing the part. They didn't really pressure me too hard, they left it up to me, but it was clear that they were disturbed, so we changed it to 'God damn you to hell'. Just because it was distracting. And one thing I prefer not to do as a translator is distract. I'm not really interested in imposing my personality in any way, shape or form, I prefer to remain invisible.

Nowadays, as I said at the beginning, rehearsals are much more of a discussion that they used to be.

RB: **And from your perspective is that good or bad?**

CH: I don't mind at all. Kristin Scott Thomas would often say, this line isn't quite comfortable for me to say and she wouldn't necessarily have a reason, it wouldn't necessarily have to do with the language, more with the sound of it and I would usually find a way to do it, which as often as not was better. If they make you think, there is no reason you shouldn't come up with something better.

RB: **It can be to do with rhythm or breathing.**

CH: Yes, it's often that or that the thought isn't quite encapsulated in the line that you've given them and I think that often these things emerge at a very late stage because it's only when the audience gets in that something doesn't get the laugh you were expecting or gets a bigger laugh or somehow gets a sort of baffled reaction. All that, you can sense when you are in the room and the actors can sense when they are on the stage. When all of this began in the late 1960s, early 1970s, there was a lot of huffing and puffing about how it's not right to pretend to translate if we don't know the language, but in fact there is no denying that the ability to write dialogue for actors is probably a more specialized or rarer skill than the ability to know what a line literally means. It seems to me that argument is over.

MP: **This was one of our questions, whether stage translators should actually have training as playwrights, work as playwrights?**

CH: I think so, I think everybody thinks that now. Michael Meyer fought a valiant rearguard action. He was the official Ibsen translator and I knew him, he was a nice man, but he was very distressed by all these people who didn't know Norwegian, including me, and eventually when I did *The Wild Duck*, he wrote a long piece in the *TLS* condemning the practice, but I think he knew that it was a losing argument. He was very good and people still do his versions, and he was a playwright, but more than being a playwright he was the Ibsen specialist, Ibsen biographer and knew more about Ibsen than anybody, and I absolutely see his point of view, but I don't think exclusivity works. When I did the first Horváth, it worked out very well, and Horváth is very, very difficult to translate because it's a kind of stylized lower middle-class language which differs from play to play. Just as Pinter has a particular language which you can recognize and which at the same time has all kinds of added resonances, so Horváth was writing in this heightened language and he said, a lot of my characters are people who normally speak dialect but I haven't written their parts in dialect, I've written them as if they were people who normally would speak dialect but are trying to speak a little better and that sets you a whole lot of conundrums. It doesn't apply to later plays like *Don Juan Comes Back From The War*, because that's written in regular High German, but the so-called *Volksstücke*, the big ones like *Tales from the Vienna Woods* and *Faith, Hope and Charity*, are written in this kind of strange invented language. Anyway at the end of doing *Tales from the Vienna Woods*, which, as I said, was a great success at the National, the Horváth Estate (he was still in copyright) offered me exclusivity on all his plays. I don't actually philosophically agree with that, I don't think you should do that to an author. There have been lots of examples of authors, for example, Pirandello, where there were various exclusivities. Brecht, Brecht notably, there were all kinds of exclusivities on Brecht translations, which don't do the playwright any favours at all. So, I'm opposed to it. It didn't stop me from being slightly irritable when three or four years ago the National revived *Tales from the Vienna Woods* and they got somebody else to do it; it was a failure. I'm not saying that it was the translation above all else, but I do think it makes a difference. The language is the frontline piece of mediation and I think if you don't get that right the play won't communicate itself. I know this from seeing my own plays in France and Germany sometimes, that audiences can be absolutely baffled. If the translation is not good, if the production is not good, the audience can't see what the play is supposed to be about.

RB: You mentioned Reza and the fact that she can turn up to rehearsals. Does she do that a lot?

CH: We do an interesting thing with her plays which interests her as much as it does me and that is that we do two versions of the plays. We do the English version and then, when it's done in New York, we do an American version. In the Faber plays anthology, *Plays 1*, Yasmina, for some reason, insisted that the *Life x 3* translation was the American version and if you compare the English version and the American version you'll see a considerable number of changes.

RB: I was reading *Life x 3* the other day and noticed it has 'Wotsits' and 'chocolate fingers'.

CH: The very first line of the play we had to change in America to 'He wants a cookie' and so it goes, all the way through. We are about to do *God of Carnage*. We start next month in New York with the American actors. What happens is that Matthew Warchus, Yasmina and I sit down with the whole American cast and go through the play with the actors, we started this with *Art*, we go through with the actors line by line and anything that doesn't work for them, we change. It's surprising, it'll run to several changes a page. We try to do it a month or six weeks ahead of the rehearsals and then they can can have time to digest it all and come to rehearsals familiar with it and familiar with the process, in case anything else comes up, which it will. Again it varies from actor to actor, some actors want every line examined and some actors . . . you work with someone like Ralph Fiennes, they will identify half a dozen places where they want a second thought, and once the previews start they'll find three or four more things that aren't quite landing with the audience in the way they want them to. The director, Matthew, plays a very important part as well, by now, both of us are very used to working with

Yasmina and her plays and we are very sensitive to the nuances. There are passages in her plays where most of the time you are just trying to deliver the goods as it were, make the lines as funny as possible, at other times, you know, the language becomes heightened and you have to find a way to do that smoothly in the English so that it doesn't stick out. And many of the discussions you have with Yasmina have to do with lines which she identifies as very important to her, which is really helpful, then you can go away and respond to them and think about them. Very occasionally she'll say 'That's not quite what I meant.' I can't remember it actually leading to her changing the French but you'll be able to isolate some nuance you can add in the English which wasn't there in the French. She's always giving these interviews where she gripes about English audiences laughing too much, but in fact there's a pretty good understanding between us, that's why the relationship has lasted this length of time. It's quite a jokey thing, the complaining that goes on. I think it goes back to a very important question we have already touched on, the question of being temperamentally comfortable with the writer you are translating. I mean, I've been asked to do lots of plays that I've turned down, not necessarily because I don't think they're good. I won't do Marivaux, I won't do Strindberg, I won't do Pirandello, because that world is not mine, so the writers that I've chosen to specialize in and translate are writers that I'm temperamentally sympathetic to and that I like to spend time with.

MP: Do you think that maybe you feel close to them as a writer, they are touching upon what you are also exploring from other perspectives?

CH: I'm sure that's right. It's a view of the world. I think all those writers . . . of course you go from Ibsen to Yasmina, it's an enormous range, but they all have a certain humour, a certain melancholy, a sceptical view, a certain dislike of self-righteousness. They are all quite impalpable things which often you rationalize after the event, you read the play and think, oh yes, I respond to this or I don't.

RB: With Reza being a contemporary author, did you feel a responsibility when you started off with *Art*, compared with Ibsen, von Horváth or Molière for example?

CH: No, actually, I felt slightly freer than I do with the great dead. First of all I knew about Yasmina before the play came on. Her play *Conversations*

After a Burial which was seven or eight years before *Art* had been found or picked out by Peggy Ramsay, my agent, who let me read it and I think Howard Davies read it too. Yasmina was then in her late twenties, but it's very hard to get theatres to do plays by unknowns, especially if they're foreign. So, seven or eight years later, I was in Paris and I happened to walk past this theatre where *Art* was playing and saw that it was by Yasmina and thought, what an interesting title, and then found that it was sold out for three months. I went, queued up, got a return ticket and thought this is really something. Then I found that the rights had already been bought by Sean Connery, but I put myself forward to do it, because I just responded so strongly to the play when I first encountered it. That's the first time that's happened to me, that I found a writer and said, I want to translate this person's plays.

RB: You say that you felt freer in terms of representing this person, but if the production hadn't worked out, that might have been difficult for her international success.

CH: I think it absolutely makes an enormous difference. She was taken aback when she saw the first preview because people laughed so much, they laughed more than they did in Paris and that's continued to be the case. I mean English audiences find her plays funnier than French audiences, I don't think it's anything that I'm doing, I feel I am maximizing the potential of the jokes, but they are there, and none of them is invented. A play like *The Philanthropist* (1970), I've seen in many different languages in different countries. National audiences all laugh in different places, it's a field of study that lies wide open. You'd have to sit in five different countries and see the same play and you'd be amazed by how the responses differ from one country to another.

RB: Looking at translated plays that you are not at all involved with, when you go to the theatre how do you judge whether you think a translation works or not?

CH: I tend to feel that it doesn't work if it draws attention to itself. I should say that I don't think there's any harm in saying, you know, this play is adapted, I mean Tom Stoppard did a couple of plays by Nestroy in the 1980s, and these are plays by Nestroy adapted by Tom Stoppard, but the label on the tin needs to be fairly clear. It is a difficulty because, generally speaking, I say 'translated by'; if it's a language I don't know, I have to say 'a new version by', but what I always try to avoid is the word 'adaptation' or 'adapted by', because I think that gives a signal that is not accurate in so far as what I'm trying to do.

RB: Do you think it implies a distance between you and the original author?

CH: No, I think if something is adapted, it's changed by definition, so, if I'm trying not to change, if I'm trying to preserve the essence of the piece, then what I'm trying to do is translating. I should say for me it's been a very, very lucky branch of my profession, it's been a very enriching thing to do and I've had a very good time with it but I think, in general, translators are very undervalued, under-appreciated, and under-rewarded. I can't say that is the case with me, because if you translate a play you get a chunk of the box office and you do very well if the play does well; but, in general, I think it's a very lonely trade in this country which it shouldn't be and isn't in other countries, where it's much more prestigious. In America it's even worse. Michael Glenny was Solzhenitsyn's translator and he was doing these massive pieces of work and dealing with a figure who was, whatever you may think of him now, incredibly culturally central and important at the time and nobody particularly thanked Michael, I don't think.

RB: There is a great deal of research on the way that translation is positioned in various cultures and Anglo-Saxon culture doesn't welcome it in the way that other European cultures do.

MP: The percentage of translated literature is very low in the UK and in the USA compared to other countries.

CH: It wasn't the case with *Art* because *Art* was phenomenal, but generally it's very hard. Thomas Bernhard, really important writer, they did a play of his at the National in 1976, *Force of Habit* and it was hated by the London critics and didn't do well and I don't think that they've ever done another play of his since.

RB: Do you think it's harder to persuade an English language audience to look at foreign plays?

CH: I have two examples of the unadventurousness of recent British audiences. When we did *Marya* in 1967, it had mixed reviews, but even though Babel was an interesting Soviet writer nobody knew much about, the run sold out immediately. When we revived it in the 1980s, when Jonathan Miller ran the Old Vic, Roger Michell directed it, it got unanimous wonderful reviews, but the theatre was half-empty. Likewise, when we first introduced Horváth, *Tales from the Vienna Woods* and *Don Juan Comes Back From The War*, which were the first two plays I did, were absolutely sold out. *Tales from the Vienna Woods* was a great commercial success for the National. When we came to do *Faith Hope and Charity*, in 1989, at the Lyric Hammersmith, very good reviews, nobody came. One of the best productions of its year was a play called *Flight* by Bulgakov at the National, half-empty. And it's very worrying that the core repertoire seems to me to have shrunk over the years so you don't get the National putting on as many interesting unknown foreign plays. You don't get the RSC doing Ostrovsky any more, you never see a play by Corneille, it's just tough to get people to take a chance on them.

RB: **So is it business rather than cultural resistance?**

CH: It's partly that. I think it's just a narrowing of adventurousness, maybe to do with the fact that if you pay so much for a theatre seat, you want a known quantity. But I think it's a great shame. These things have their cycles . . . I think the fact we used to have the World Theatre Season where you could go and see plays, you could go and see a Spanish company doing *Yerma* by Lorca, an Italian company doing *La Lupa* by Verga with Anna Magnani. Every year, interesting plays would come from other countries and people were quite accustomed to seeing those plays. It doesn't happen now. It's very rare now that a production in another language comes to this country.

RB: **Do you think there is a difference between the UK and the US?**

CH: America is even less adventurous. It's the same argument as when you say the rich get richer and the poor get poorer, there'll be three productions of *The Seagull* in a year and poor old Horváth won't get a production from one year to the next or one decade to the next. I was very conscious of this, because in the 1970s my plays were done a lot in Germany, usually in many productions. I had a very good translator, a novelist called Martin Walser, who also wrote plays and a play of his was done by Charles Marowitz, at the Open Space on the Tottenham Court Road, a play that had had great success in Germany, and Martin came over, we went to see the play and there were ten people in the audience, which was very embarrassing to me, because I would go to these vast municipal theatres, 1500 people coming to see your play. It was just a sign of how insular we are.

RB: **One of the contributors to the volume, former National Theatre Literary Manager Jack Bradley, notes how difficult it is to promote what's not in the repertoire.**

CH: Yes, well, finances are so precarious and obviously it won't improve in the next few years, but I think it makes the work of the translator doubly important really. To try to make available the byways as well as the main arteries of culture.

Note

1. The dates given in brackets indicate the first publication of the relevant translation/adaptation/play by Christopher Hampton.

11
Not Lost in Translation[1]

Jack Bradley

From 1994 to 2007 I was Literary Manager at the National Theatre on the South Bank. As such, I was over time advisor on programming to Richard Eyre in his later years as Artistic Director, for all of Trevor Nunn's tenure and, most recently, assisted Nick Hytner during his preparatory phase and his incoming years. I was not, of course, sole advisor; each regime has had a slightly different structure and consequently different decision-making process involving to a greater or lesser degree other advisors. Moreover, obviously regimes shift and change according to need and the natural evolution of the personnel involved. However, it is possible to detect certain common threads within each management and, consequently, to derive a sense of how the building has approached translation and adaptation in recent years: in essence, the implicit policy to foreign work, the processes by which work is commissioned or chosen. I would like briefly to look first at the period when I was working on the South Bank and then to widen it out to take a snapshot of work firstly in the West End now and elsewhere in the country in the hope that we can determine whether there is a discernible national policy or approach to foreign work. In the simplest terms: how interested, as a nation, are we in what other people, nations and cultures have to say? Where does inclusivity begin? At our borders? Or further afield? In an increasingly shrinking world, do the neighbours of the global village really understand one another and what are the arts institutions doing to enhance that understanding?

But before I begin to address some of these issues, I should perhaps tell you a little about the theatre in question. The National Theatre of Great Britain (NT) is situated on the South Bank of the Thames in London, almost opposite the Houses of Parliament. Built just over thirty years ago, it was the product of a campaign that took almost a century to succeed. Obviously, there was no shortage of theatres in the West End, which abounds with cramped – if rather decorous – playhouses from yesteryear, now largely dedicated to twentieth-century revivals, blockbuster musicals and the occasional transfer of a new play from the state-subsidized sector. The nation – or at least a small but vocal minority, including G. B. Shaw and Harley Granville-Barker, wanted something more: a theatre that would celebrate dramatic art, offering an international and classical repertoire, an institution that placed the writer at the centre of the creative process. And so the long campaign began to create a site to make the dream concrete. And concrete it indeed proved to be. Finally completed in 1976, it was designed by Denys Lasdun and sits imposingly, its grey breeze-block and angular lines a provocation to almost all of the classicist architecture north of the Thames.

If it invites opinion – and it certainly has done, with some going as far as to compare it with the brutish facelessness associated with nuclear power stations and other ominous buildings of dubious utilitarian value, few do not warm to it on entering. The large and airy ground floor lounge is a hive of quiet industry throughout the day as theatre makers from all over informally use it as a convenient meeting place. It is not uncommon, walking through, to see a young hopeful spreading their design portfolio before an expectant director at one table, whilst in another corner a bevy of angry fringe artists mutter darkly about their bourgeois surroundings over a soothing cappuccino. But if the building is a lodestone for the industry, it is, of course really defined by its theatres and the work you find within them.

I will dwell a little on their respective designs because each is very particular to itself and to the plays that work within them. When it was first conceived, there were to be two stages at the NT: the Olivier and the Lyttleton; however on its completion it was discovered that to the rear and side of the building, near the scene dock, there was a 'dead space', an area not designated for building or storage. People realized that with a little more money this could be turned into an entirely flexible, black-box studio theatre, and so the 300-seat Cottesloe – an intimate, by comparison to its two sister theatres – came into being. And as so often happens, this afterthought, this last-born has come to be most loved. Loved by writers because it allows for small details: they can make noise quietly as it were; loved by actors, confident their subtleties will not be lost on their audience; and loved by directors because its movable seating means countless configurations are possible.

In sharp contrast, the other two theatres have comparatively much less leeway for change. The largest of the three stages is the Olivier.

Its auditorium seats over 1000 people. Moreover, the stage design is positively classical. Imagine a Greek amphitheatre, but with a roof on it: a huge apron stage thrust out to meet its audience. It is something like 28 metres from the prompt corner to the centre stage where an actor can take in the fanned arc of seating that is a full 162°. Naturally only certain kinds of plays relish such a playing space. There is a rule of thumb: you can do a big play in a small place – in fact we once did *War and Peace* in the Cottesloe to prove it! – but you can't do a small play in a big space. Somehow the ideas and the *mise en scène* gets 'lost' in the seeming cavern of such an empty space. Actually, I suspect there is an exception to this: you could, I imagine, do *Waiting for Godot*, where the cosmic sense of isolation could be enhanced by playing Vladimir and Estragon in such a vacuum. But such plays are rare and instead history has found the Olivier more naturally welcoming to certain kinds of plays. Not surprisingly, plays that use declamatory styles or direct address, such as *The Bacchae* and other Greek classics; modern celebratory plays (I'm thinking here of works that utilize classic music scores like 'Guys and Dolls') or plays that portray whole communities – such as Shakespeare's and what I call 'public' plays, such as Ibsen's *Enemy of the People* (which is also, if you think about it, a community play) or more recently the overtly political plays of David Hare which are specifically designed to question our public institutions: the press in *Pravda* the church in *Racing Demon*; politics in *Absence of War* and the causes of the Iraq war in *Stuff Happens*.

Which brings me to the Lyttleton. In some respects, a traditional proscenium-arch theatre resembling in principle the antique playhouses of the West End, but this 900 seat auditorium is perhaps twice as large and twice as wide as many of its eighteenth-century counterparts across the river. (In fact, critics in the building occasionally refer to it as the multi-plex because of the flattening effect of this modern proscenium arch.) Once again we find ourselves asking what kind of play will sit up in such a space: what kind of rhetoric will 'hit the back wall' and involve an audience entirely? Ibsen and Chekhov are popular there, though you should not presume such classical conservation rules absolutely: in 2002 the Lyttleton played host to the new through-sung modern and wonderfully tasteless *Jerry Springer: the Opera*. Of course, with programming such as this and the dance/movement piece *Play Without Words* by Matthew Bourne, you could say that with the arrival of the twenty-first century, the NT is consciously trying to dispel the repertoire assumptions that have evolved over the last thirty years. As a national institution, it has a custodial responsibility to provide every new generation with access to the classics but it must also never become culture in aspic. New forms and new voices must also be embraced.

I have focused upon the design of both building and theatres in an attempt to convey the ambience and the mission it perceives itself to have: the challenges and opportunities that come with being theatre's flagship in England. The last element to touch upon is the volume of the work. As I have said, I was at the NT for almost thirteen years. In that time, I have seen around 250 shows open in the building. That is not unusual: on average the NT makes almost twenty shows a year and in addition may also receive international visitors in the same period. That means given the need to forward plan, you find yourselves every two or three weeks having to ask: what are we going to do next? Even though it won't be seen on the stages for a further nine months, it is still the pressing question of the day. When you realize that the NT is in daily competition with the RSC (Royal Shakespeare Company) and all of the West End, the imperative to find the new, to revive the forgotten and unearth the undiscovered, is a sizeable task. It requires exhaustive research. I once described the dilemma as having to think of something everyone had heard of which no one else had yet thought of doing. Moreover, unlike many theatres in Europe, it does not close in the summer. In fact, it never closes, except on Sundays. And, indeed that is about to change this autumn with selected openings on Sundays.

I have dwelt upon the architecture of the respective theatres at length for I hope good reason. As we turn to the choosing of the actual repertoire, it is impossible to disregard both the nature of the auditoria as well as the economics of running the organization. With public funding making up £16+ million, but with the theatre's annual turnover touching £50 million, then Box Office income remains a major consideration in both programming terms and general viability. In practical terms, that means hitting targets of around 70 per cent or 1600 tickets sold daily (more on matinée days). The repertory system is a cushion of sorts. It enables you to interweave the popular with the esoteric to create a balanced and appealing programme, but there is no getting away from the fact that if a show does not take off, then the theatre may find itself staring at a sizeable period of pre-announced performances for which it cannot give away the tickets. A disaster in the larger auditoria means the budget can haemorrhage over a million in a few weeks. There is a perpetual tension between the temptation to be conservative and the wish to be daring. The question is which impulse wins out.

For the answer, let us take a look at the years 1995–2006 inclusively. As I said, I watched around 250 shows premiere during that time. They can be categorized in several ways, but it is helpful at the outset to remind ourselves that the historic aspirations of the National were to provide an international, classical and contemporary repertoire. Or, in other words: both the old and new and drawn from everywhere. To this end my predecessor, Kenneth Tynan, assembled a list, entitled simply, SOME PLAYS. It is a list, nothing more, not a digest nor story synopses collated to assist Associates to choose their work from. It is a list which is chronological and geographic, from Aeschylus through the Romans and medieval times, becoming increasingly fulsome with the Renaissance before touching upon the Spanish, and Italian golden ages and the Romantic period. Unsurprisingly, it finishes post-war. It is not complete nor is it impartial. Not all of Noel Coward is included and, elsewhere, there are strange omissions – no mention of von Horváth, for example. The evidence suggests it is an assembly of titles from secondary sources, only. It was, however, in my first years on the South Bank something of a bible until I started spotting the omissions. However, despite its shortcomings it was testament that the aesthetic imperative laid down by Granville-Barker et al. had not been forgotten. But to what extent was it consulted?

If one analyses the 250 plays you find that they breakdown approximately as follows:

Shakespeare 21
Musicals 16
Adaptations 21
Revivals of British or English speaking world classics 48
Foreign titles 41
New work or new plays 100

Of course, numbers in all their baldness tell us little. The Shakespeares vary from a sell-out but limited run (approx forty performances) of *Anthony and Cleopatra* starring Helen Mirren (she was too busy to give us an open commitment) to world tours of Sam Mendes' *Othello* and Simon Russell Beale's *Hamlet*. Not forgetting a couple of performances of *Shakespeare Unplugged* by the Education Department in the Cottesloe.

So the numbers mislead, especially in the case of new work and plays. Ken Campbell's *Theatre Stories* (eleven performances) counts as one of the 100, so does Marber's *Dealer's Choice* (thirty-five performances) as does Alan Bennett's *History Boys* (250 performances and counting, not including Broadway or national touring).

Equally, should *His Dark Materials* (300 perfs. approx) count as two because it is in two parts? Or *Guys and Dolls* (120); *Little Night Music*

(100) ditto *South Pacific, Oklahoma, Anything Goes* or *My Fair Lady* count as only six shows even though they each ran for many months? In other words, shows do not punch with the same weight. By scale of numbers certain titles have a greater cultural impact.

Likewise, the revivals of 'neglected' British classics dominate beyond their supposed 20 per cent of the repertoire. *Wild Oats, London Cuckolds, The Relapse* and *Money* by Bulwer-Lytton were all popular hits and garnered runs of around 75–100 performances in the main houses. Significantly, they are all comedies, albeit one, *Money,* has slightly more satiric edge. In other words, the core audience of the National have systemically voted with their feet for English comedy and American musicals. In fact, if you combine American musicals with twentieth-century American plays, thanks largely to Miller and Williams, the number of American titles produced during the period rises to twenty and again, these have, by and large, proved very popular with the Box Office.

So how has the rest of the world fared? Let us take a look at the forty-one plays translated from a foreign language during that period:

Ancient Greek 9
Russian 9
German 9
Norwegian 4
French 8
Swedish 1
Italian 1
Dutch 1
Australian English 3

The ancient Greeks seem to fare well, though they are slightly flattered by treble and double counting respectively the *Oresteia* and *Oedipus* plays. The Russians are rather reliant upon Chekhov (four productions) for their place in the tables as are the German-speaking nations, since Brecht makes an equally strong showing. France have the benefit of two Cyranos (one re-located in India by Tara Theatre) two Marivaux and, despite the emergence of Jon Fosse during this period, Ibsen accounts for the entire Norwegian presence. Strindberg would not be pleased to be ousted by his arch rival, nor the Italian showman De Filippo particularly impressed. However, there is one striking fact: during this period of twelve years, there is only one play produced written in a foreign language by a living writer: *Life X 3* by Yazmina Reza. (You may increase this to two if you include Mary Morris' adaptation of the novel *Two Weeks with the Queen*.)

I will return to this arresting fact, but firstly, is this a fair snap shot? If compared with a twelve-year period of the late seventies–late eighties, that is to say, under Sir Peter Hall, one finds that there were forty-three foreign works in a period of 190 shows. This implies that Hall's taste was more pan-European; however, there are a number of other factors to take into consideration. For example, in 2002 as part of the Transformation Season, the NT created a fourth space known as the Loft where additional productions occurred. The volume of new work increased dramatically that year. Likewise, as the newly arrived Artistic Director of the NT on the South Bank, Hall's earlier remit was to present international titles previously unseen or barely know in the UK. His successors are at some disadvantage in that in terms of presenting the classical canon, they face a law of diminishing returns. Once a title has been premiered, the sheen of novelty is dimmed and tarnished. That isn't to say, one cannot replicate titles – indeed, as part of his first £10 season, Nick Hytner produced Horváth's *Tales from the Vienna Woods*, not least because it had not been seen since the historic production almost twenty years earlier at the Olivier. Last year, *Galileo* was revisited for the first time since Gambon did it twenty years ago. Indeed, as a rule of thumb, one begins to detect a trend: a landmark production of a title effectively puts that play beyond the pale for a generation. Of course, Shakespeare confounds this assumption: Hamlets can come and go (indeed, I can recall three Lears in a single calendar year) but tellingly, in the same twenty-year period, I know of only three *Don Carlos* productions and as many high profile *Mary Stuart*s in the same period.

And what of the rival National company the RSC? Their remit is obviously very specific, but their dedication to the bard has been tempered by a commitment to new work and, in recent years, the presentation of Shakespeare's contemporaries and themed seasons such as the Spanish Golden Age. The ratio of foreign work in their repertoire is similar to the NT and, with the notable exception of a Koltès, they have not been noted for premiering work by foreign contemporaries. It would seem the companies mirror one another in their approach to international work.

To complete the circle, I should like to offer a snapshot of the West End (including the subsidized sector) and fringe theatre from June 2007. Lifting a listings magazine, it is all too easy to see the scope and variety available. Or lack of it. Of the 55 theatres in the commercial (and subsidized) sectors, twenty-five are offering musicals; five are offering such spoof adaptations as *The 39 Steps* and *The Hound of the Baskervilles*; there are eight revivals of, again, largely twentieth-century British classics such as *The Letter* or Pinter's *Betrayal*. The inclusion of The Globe and the Regents' Park Theatre means there are six Shakespeares available and dotted around the capital ten new plays to choose from. There is one foreign work: *Kean*, by Sartre, though significantly a play based not on a French performer but a famous British actor, starring Anthony Sher. The only light at the end of this ever-narrowing tunnel is the imminent arrival of *Elling*, an adaptation of an award-winning film from Norway that transfers from the Bush Theatre after a sell-out run. It stars a first-rate TV actor whose show *Life on Mars* won plaudits only months ago.

On a brighter note, the fringe itself shows more daring, as you would expect. The Bush is hosting a contemporary Japanese play whilst small-scale productions of Sartre, Camus and Ibsen litter Zone 2. It can be summarized thus: forty venues, five musicals, five foreign works, four revivals and the rest are new works of some kind.

The message is clear. The British Theatre scene is not lost in translation. Its audiences revel in the familiar – plays they know – songs they can sing along to, or new plays on subjects that are close to home, that somehow seem more pertinent to them. The trouble is I find this paradoxical. Why when we celebrate culinary diversity, when we thrive on foreign travel until our carbon footprint is charred, do we seem to prefer to curl up by our TV-warmed hearth with a meat and two-veg diet? What can we do about it? And how has it come about?

Lifting the lid on play selection

Well, the how has it come about is a little easier to identify. Let me talk about custom and practice, the Realpolitick of choosing plays. It is often said in England that British Theatre is a writers' theatre, that the playwright is paramount, their voice sovereign. It is true that in theatres dedicated to new work, there is a palpable hunger – almost obsession – for the new and the latest voice. If this sprang from 1956 and Osborne's impact on Sloane Square, the ripple effect has, over the last 50 years, spread out from SW1 throughout the London area and into the regions. When I began in Literary Management twenty years ago, very few theatres had full-time Literary Managers, now most have Literary Departments. Inevitably, this has an impact on the programming of work, as buildings vie for the next new thing. This has its own effect: the successful profiling of playwriting leads to more writers trying their hand. It becomes the art form de nos jours. If in the late eighties Granta could publish – and did – articles and lists entitled '20 Novelists and 20 poets to watch', since the nineties, thanks to the notoriety of such writers as Sarah Kane and Mark Ravenhill the attention, to some extent, has returned to the theatre. As Rosie Cobbe, agent at United Artists put it recently 'we are in the export business'. She was at the time explaining why writers could not rely on second productions in Britain to make a living, but she also, incidentally, summarized the cultural assumptions in Europe and beyond. If we are in the export business, then as far as Germany were concerned in the nineties they were in the import business mopping up the works of the blood and sperm brigade, as a German journalist so colourfully put it.

Actually, it makes for a curious dynamic. At a conference a few years ago in Novi Sad, myself and Graham Whybrow, then of the Royal Court, were berated by a Croatian delegation for corrupting their youth with Brutalist British theatre. The curiosity for me lay in the fact that we were guilty for premiering these works, but they took no responsibility for choosing to translate these runaway hits. It would seem the cult of the new that colours decision-making in British Theatre is as alive and well abroad, the difference being that foreign dramaturges and directors seem, in the first instance, more receptive to work emanating from Great Britain. Actually, setting aside cultural reservations about the subject matter of British plays, there are sound reasons for this, namely: there has not been the same proliferation of playwriting in Europe as we have seen in the British Isles over the last 50 years. Whether it is because of the stultifying effects of Soviet repression and censorship and the concomitant need to conceal intent in symbolism and more oblique forms of theatre, the effect was to drive many writers from the theatre. As a medium, it became in many cases moribund. The net effect: my colleagues in Hungary, Poland and in the Balkans would concede that they still only receive 50–60 plays per annum at best. The Royal Court routinely receives 3000. The sums speak for themselves.

But the sums tell another story. I began this section by saying that in Britain it was a writers' theatre. In fact, it's not. Never has been. It belongs to directors and they put on plays that they like. What they like is determined by what they have read and if they have less exposure to foreign work, they will without thinking retreat into a comfort zone of familiarity, just like the majority of audiences. It is further complicated by two things. With some exceptions, the nature of being a director is not to read widely, but to read the same play over and over again as you struggle to find a way of doing it. This is more than laudable, it is essential. But it impacts on one's time. How wide can one's reading be in such circumstances?

The second thing is the capacity to read and discover foreign work. There is the old British Council joke: what do you call someone who speaks three languages? Answer: tri-lingual. And two languages: bi-lingual. And one language: English. Our education system, if not ourselves, does us no favours in failing to embrace the value of language learning. But then value is a slippery creature. I predict our children will be encouraged to learn Mandarin or Hindi before long. After all, in the modern day, if it makes business sense, it makes the only sense that matters. Don't get me wrong. Before I left the National I had been lobbying for a Chinese play – *Providence of the Heart* – to be done, though whether knowledge of a sixteenth-century Chinese love story is quite what Big Business has in mind, I doubt it.

So how does foreign work get produced, facing as it does cultural and economic conservatism, an ignorance of contemporary and classical foreign work and an inability to read it in the original and a home grown market that is bursting at the seams? The answer, in my experience, is a combination of stealth and guile. I will give a slightly left field example and then an application of the same thinking. In the late nineties, I did some research on the back of an envelope that confirmed an impression. In almost twenty-five years, there had only been six plays by writers of colour presented at the National Theatre. And even then, it required including Ranjit Bolt translating Brecht to bump up the numbers. I didn't think it was good enough. Trevor Nunn was dismayed. As it happened, via the NT Studio, I had routinely been working with emerging black British writers such as Tanika Gupta, Roy Williams, Ashmeed Sohoye and so on whose work then went to be produced elsewhere in the capital. We upped the ante and invested more. The outcome: main house productions for Roy Williams, Tanika Gupta, Kwame Kwei-Armah, and an incoming production of the *Ramayana* from Birmingham and so on. When I left we were averaging almost two shows a year from writers of the black and other minority communities. Most recently – and symbolically important – Nick Hytner directed Ayub Khan Din's *Rafta, Rafta* in the Lyttelton. Not exactly cause for complacency, but a modest start. The assumption now is that the programme must include work from under-represented communities. That was a slightly left field example of gently imposing a programming strategy designed to impact on the diversity of the output.

As it happens, likewise via the Studio, there was an initiative to expand our translation work. However, in the first instance, it could hardly be described as strategic. The truth is as part of the French Theatre Season, we had been given monies to pay for a workshop to be conducted by Mnouchkine. Several attempts to bring her to London had faltered given her other commitments. The momentum was ebbing. We would soon have to give the money back, unless we came up with another project. No one likes giving grants back so we put our heads together.

I had become aware that translation work in the British Theatre scene was more or less sewn up. The same few people were increasingly asked to do the relatively few available high-profile translations. There was nothing sinister in this: people like to work with those they have a working understanding with. However, to be brutally honest, we needed to create a new generation of translators. (Before we go any further, I should say that the NT policy was, though not exclusively so, one of using playwrights equipped with literal translations to make versions of the original. I realize that is not universally admired but it was the custom and practice of the time.) We enlisted Philippe Le Moine, as the name suggests, a Frenchman, and we concocted a scheme called Channels which allowed us to identify five French contemporary plays which were then translated and received rehearsed readings at the NT as part of the Transformation Season. The work was more than simply translation. The British playwrights received a workshop on the difficulties of translating from their past masters who generously shared warnings of impending elephant traps. Moreover, the French and British playwrights spent time together and so there was a modest degree of cultural exchange, overseen by Philippe and their respective literal translator. Over subsequent years, similar projects went ahead with exchanges with Hungary and Argentina and Channels gradually morphed into Inter-text, a Europe-wide translation collaboration between France, Austria, Italy and Germany. To date, none of the work has been shown in a full production at the South Bank but there have been regional productions at Birmingham, Glasgow, at the Gate in London and on BBC Radio 3 to name but a few.

The impact then has been strategic, but low profile. A much more public response to internationalism has been the work at the Royal Court under Elyse Dodgson over the same period, though interestingly ten years ago, their first German season comprised a week of readings and discussions in the Theatre Upstairs. Shortly after, when there was a 'gap' downstairs that needed filling, people remembered enjoying one of the readings and so the play was, to use an airline term, upgraded. Over the decade the international work has contributed more and more to the fabric of the Royal Court programming, vying alongside the Young Writers' Programme for space at Sloane Square. We have had full productions of plays from as far afield as Russia and Mexico. The stranglehold of plays about Barons Court bedsitter-land seems to have been broken. Some small progress has been made. Indeed, it is not so very long ago that Kroetz found a revival of his in the West End, something few would have predicted a few years ago.

I promised you stealth and guile. By that I did not mean simply the softly, softly approach of the NT Studio or the Royal Court: it was a coded reference to the ways Literary Managers and dramaturges attempt to land a production. When I came across *Fair Ladies at a Game of Poem Cards* by Peter Oswald based on the puppet play by Chikamatsu, I decided quite carefully when to suggest Richard Eyre read it. Richard had once described running the National as eleven years of tooth-ache. I knew the further he was away from it, the more fondly he thought about it. So, as he planned to board a flight to New York, one of his favourite cities, to direct at the Lincoln Centre, one of his favourite theatres, I handed him Peter's (almost unsolicited) play. 'Something for the plane'. Richard was in a good mood. He glanced at the cover. 'Do you know', he said, 'Peggy Ramsay always said I should do a Chikamatsu before I finished.' With that I could see him gently taking ownership of the idea of the play. I had a hunch when he returned we would programme it.

Literary Managers are advisors, advocates, occasionally gate-keepers and cultural bodyguards for their building. It is never they that have the last word, especially if the Artistic Director is passionate to present a play or an Associate has persuaded that Artistic Director of a play's brilliance or importance. The NT did *La Grande Magia* I suspect as much because we had all heard of and read about Strehler's production at the Piccolo. That did not mean we wished to replicate it, simply that we had been assured by its success that the play offered a great night in the theatre. I like to think after we did that play other practitioners remembered *Napoli Millionaria* as fondly and were encouraged to go off and read other works by De Filippo.

I have to confess sometimes less honourable motives drive us. Whilst thinking about rare Russian plays for Declan Donellan, I recalled Erdman's *The Mandate* and *The Suicide* (recently riotously re-vivified by Moira Buffini at the Almeida). I should have recommended *The Suicide*. It is a better play. But I didn't. I recommended *The Mandate*. Why? Because I had seen *The Suicide* when the RSC did it twenty years earlier at the Aldwych season. I hadn't ever seen *The Mandate*. I wanted to see this other play. Of course, I justified it to myself by saying that the critics might remember the RSC production and compare them. I could argue that my choice helped to enlarge the British knowledge of the classical canon, a worthy aim in itself. Truth is, I was, in a quiet way, showing off my erudition. Just as I berate actors who move through their career ticking off the classical roles, and directors who likewise are drawn to major texts because the great play offers depth, challenge and the security of robustness, I am guilty of my own prejudices and objectives: I like watching what I've never seen before. I also like showing off.

I came late to foreign work. My first passion was new British plays. It's what I and my friends wrote, it was what I understood. Slowly, by looking further afield, I became a convert, albeit it feels that my evangelism seems to have made only a small difference. I have a long list of lesser known works – classics elsewhere – I would have loved to have seen done, but I didn't get it all my own way. Yet it is important the campaign for real inclusivity continues; if, as Shakespeare said, death is another country, then I now believe other countries are as much our cultural life-blood as the English classical canon or the latest new thing to burst onto a stage in Sloane Square. We ignore foreign work – new and old – at our peril, if we are to understand our neighbours in the global village.

Note

1. This is the text of a presentation Jack Bradley gave at the 'Staging Translated Plays: Adaptation, Translation and Multimediality' conference held at the University of East Anglia in Norwich (UK) in July 2007.

12
Roundtable on Collaborative Theatre Translation Projects: Experiences and Perspectives

Jonathan Meth, Katherine Mendelsohn, Zoë Svendsen

> This is a transcript of a roundtable which took place on Saturday 30 June 2007 as part of the 'Staging Translated Plays: Adaptation, Translation and Multimediality' conference held at the University of East Anglia Drama Studio in Norwich, UK. The roundtable was conceived as a spontaneous discussion between practitioners working in the field of international drama, without access to prepared papers.

MP – Manuela Perteghella (Chair)
ZS – Zoë Svendsen (Independent theatre maker and researcher)
KM – Katherine Mendelsohn (Literary manager, Traverse Theatre, Edinburgh)
JM – Johnathan Meth (Director writernet)

JM: I'm the director of writernet and I run a network called The Fence which gave rise to a one-year Culture 2000-funded project called Janus. Writernet is a national organization that has been going for around twenty years and exists to support playwrights and the ecology in which playwriting happens in the UK, and it does this in a variety of different ways. It has a website[1] with advice and guidance for playwrights and those who make playwriting happen; we develop different projects that help playwrights and sometimes co-produce plays, and we run a series of networks of playwrights regionally, nationally and internationally and sometimes we work with cultural operators. What I mean by cultural operators is a blanket term for those working, sometimes in academia, sometimes in the profession, sometimes in the ministries, sometimes in information centres, the wide ecology of people who are involved in making new dramatic writing happen. Three and a half years ago I set up a European network for playwrights and cultural operators called The Fence. There is an idea that is finally beginning to gain more currency among the practitioners in the UK, that actually internationalizing our practice might be a good thing as artists and, in this case, as playwrights. I managed to persuade funders that my idea for a European network for playwrights and cultural operators could sit quite comfortably in a portfolio of cultural activities. I talked to a dozen different people from the UK and abroad and asked them for recommendations for playwrights and cultural operators, to invite them to a week that would be split into two, the first part of the week would be an opportunity for us to meet, and we went to John Osborne's old house that is now an Avon Foundation Writer centre in Shropshire, on the Welsh border, and then the IETM (Informal European Theatre Meeting) was coming to Birmingham. We split the week, enabling some people from continental Europe to meet their UK counterparts in Shropshire and then went to meet the 500/600 people who were in Birmingham. So we began a dialogue that's been going on for the past three years and sometimes we meet across Europe, we meet in an ancillary way to the IETM. Out of the network came the idea of a one-year project, for which we would apply to the Culture 2000 pot. We came up with something around the themes of cultural identity and cultural diversity. I had already looked at the UK cultural priority policy which is called cultural diversity and the European cultural policy is called instead cultural mobility and gone 'cultural diversity, cultural mobility, aren't they the same thing, in practice, when you get down to it?' Isn't it about how you negotiate difference, whether it's a literal border or a metaphorical one, whether it is age, gender, ethnicity, infrastructure, it's just different people. We managed to put together a winning narrative. We had the network already constructed and we put the network to work to create fifteen plays from a shortlist of 100 plays, then we had some fantastically rough and tumble arguments about what the process would actually be, often split into playwrights and cultural operators camps, but we worked it all out and recognized that the process of recommendation from the network would have to be measured against the reality of where these fifteen plays were going to go, to have their showcase readings at three festivals in Leeds in the UK, in Tampere in Finland and in Graz in Austria and we came up with our fifteen plays, for the Janus project. Two of the people I talked to, to get advice from about the project, were Katherine Mendelsohn and Philippe Le Moine (who'd been running the Channels project for the Royal National Theatre Studio), and when it came to be Leeds' turn, and West Yorkshire Playhouse who are our Leeds-based partners for that leg of the Janus project, they were quite interested in looking at that model that had been developed through the Channels project.[2] Not surprisingly as directors of an organization that puts the playwright at the centre of its work, we managed to get quite a lot of writer-to-writer involvement which meant writers usually coming across to the host country where the festival was with a week to spend working with their partner playwright as it were. Lots of things worked in interesting ways, but probably, some of the most interesting stuff came from what didn't work. For example, our idea that you would have a playwright-to-playwright encounter somehow mediated through some kind of literal translation was profoundly challenged by our Czech playwright and the translator he brought with him. It was problematic and difficult, and painful but actually a year later, we stepped back from it and we understood why that model didn't work for them and we needed to either exclude him from the project or adapt our practice. The Czech Republic has existed as a small nation, in the middle of the European continent, preserving its language, as opposed to Britain, on the edge of the continent, sharing the world's dominant language – so the position of translation is central to Czech culture in a way that it is effectively peripheral to our own. Personalities aside, it was perhaps inevitable that our playwright-to-playwright model, with the use of literals, was not going to fit.

KM: I'm the Literary Manager of the Traverse in Edinburgh, a theatre that almost exclusively produces the work of new writers in its current incarnation. The Traverse has been around for forty years and it's evolved under different artistic directors, but I would say that particularly in the last twenty years there has been a strong focus on living writers. So we don't do plays that are revivals and most of the work we do is world premieres and comes from commissions from writers. I am very passionate about international work and having a chance to see it on our stages in productions rather than readings. I think that in Britain we do a lot of readings because it's expedient and a lot cheaper and if you have a limited number of slots for production, you can't put on a million productions, much as you would love to, but my interest in international theatre probably comes from the fact that my parents are not originally from Britain, which gave me early exposure to a lot of international work. I think the personal is a good way to start, it's where we all start whether you like it or not and for me it meant that I had a very strong interest in international work. Before I came to the Traverse,

I had originally trained as an Assistant Director and worked for companies as diverse as the Royal Court Theatre, and the Royal Shakespeare Company, and then I became Literary Manager at the Gate for the first time with Artistic Director Mick Gordon and producer Philippe Le Moine. There are fashions in international theatre and at the time the fashion wasn't necessarily much for international theatre, particularly foreign contemporary work, and the Gate was the one place in London where you really could programme a whole year of international work in translation and that was a really rare thing. And also you didn't have some of the box office pressures that I have at the Traverse, because the Gate is a hundred-seater theatre and its audience expected us to programme quite radical work. The Gate allowed me to indulge myself in that enjoyment of international theatre and also to work with Mick and Philippe because Philippe had a genuine commitment, passion and enthusiasm for international work and Mick similarly has worked in many, many different countries directing work, it was a good combination of three very different people. From that, I went to the Traverse initially to do the international work and my title used to make everyone do a double take or laugh, because I was called International Literary Associate, and this just seemed a bewildering title which didn't describe what I was doing, so the way I was able to explain it to people was that it was about import and export, but of texts and writers rather than of productions. It was about literary management, it was about bringing some international contemporary plays to the UK not in their original productions, but selecting the international play-texts that you would commission translations of to present as part of your season and conversely working with the Scottish writers and their plays, and theatres abroad so that their work was translated into other languages. I was helping facilitate projects for theatres abroad that wanted to do a season of British plays or have the UK playwrights there to discuss their work. It's very interesting working with contemporary writers in translation; you have a very different possibility in terms of collaboration, you have the writer there, you can, they are living, it's not like when you have a dead writer and you are looking at lots of references and making guesses, even if they are very educated guesses, on what the intention was or the aim or the main important themes of the play are. With a contemporary writer they are there and they can speak for themselves. The way that we usually work at the Traverse is that we have the original British author in rehearsal so when working on international plays there was no reason why we shouldn't be doing the same thing. So my job in those early days was import and export of play-texts, and it was really about how do you get that work onto the main stages, how can you make it not a specialist thing but something that any audience member wants to go and see. It was interesting that after working in London before coming to Edinburgh I was aware that there was often a funny distinction between what was considered foreign and what wasn't. Somehow Chekhov and Ibsen have become British, they are not 'foreign'. Theatres in the UK programmed Yasmina Reza's *Art* which wasn't seen as foreign, only because it had been in the West End, had been attended, and was popular. Somehow there was a (wrongful) distinction seen between foreign and popular and I was very interested in how to minimize that gap. The thing about initiatives like The Fence is that you work with ten playwrights from a country, whereas if you are working on one production you'll be able to focus on just one playwright from Finland because you loved that play and you feel it has a connection for your audience, and the director. One of the joys of the last eight years is that I have been able to develop our work a lot to work with countries from much further afield than our European neighbours – such as countries from Asia and Africa – in very different ways, but what that means is that you have even more choice for your few international productions. The way of working that I developed, was a way to encourage an overlap of audiences, that is that people would come to a play not because it was from a certain country or because it was their specialist area, but because they thought it sounded like a good night out, a play worth seeing. The way I chose to do that was to develop PLAYWRIGHTS IN PARTNERSHIP – a way of working that was quite similar to what Philippe Le Moine ended up doing at the National. My project was about involving the original playwright very heavily in the translation of their own work and having them there to really probe the textual qualities, the meanings and issues but also the subtextual things, the energy of the rhythm, all of those things that every good stage translator is used to working with. I managed to get funding to enable us to commission new UK translations of contemporary plays with quite substantial writer/translator residencies and development time attached and it enabled us to commission three new international plays a year for the first three years. So far 100 per cent of them have been produced which is the one achievement for which I think the Traverse ought to be really proud as a theatre. I don't make the call as to what gets the money for production, so the programming for production is about the theatre and the Artistic Director and their commitment to that international work. We often work on achieving the plays' chances of production by matching a foreign playwright with a contemporary British playwright to work together to produce the UK English language text. But it's not about the British playwright doing something superficial like using a few slang words and saying 'oh, I'll set it in the home counties' (which of course in Scotland would not be popular!), it's rather about them *really* understanding that play because then they'll know that (for example) to use 'shag' in a context might be the wrong word, maybe there's a much more accurate word. That's the way that the very best translators work, it's not anything new. I think David Johnston once observed that translators are writers, whether they have written plays or not. For me, the best stage translators are writers in this sense, and in the case of PLAYWRIGHTS IN PARTNERSHIP, we use actual playwrights and it has worked well to achieve many of our initial aims which were partly about targeting audiences as well, because an audience is tricky to attract to a play by a writer from Finland who they've never heard of, but if you also have David Harrower or David Greig involved, then they might give that play a go. But it's about breaking down the fear barriers, not about hiding where the play is from. You must get an audience in the door first, but don't mask where the play is coming from. So we've never 'relocated' international plays in translation.

ZS: What I'm going to talk about is slightly different in the sense that I look at the Gate Theatre in Notting Hill, London, as a project, rather than looking at an individual taking a particular route through dealing with questions of translation. I'm going to give an overview of what the Gate is and then what I think are the two main elements of translation policies. I will then conclude by talking about stages of translation as I understand them. The first thing to say about the Gate, which started in 1979, is that on the one hand, almost from the beginning, it programmed *home-grown* international theatre, that is, plays not only in translation, sometimes English language plays, but produced, directed by British people, theatre makers and on the other hand, it has produced many of the foremost theatre makers of the past twenty-five years. So there is an interesting relationship perhaps between those two things. First, it's a room above a pub, which may also say something about the reception of international theatre in England as it's the only English theatre that continuously champions the translation and production of foreign language plays. Second, it's always a collaborative process, translation at the Gate, and often it is a collaboration between ideals and pragmatics, between the desire of the people making the work and the time available to actually get the work on stage. We're talking here about a four-week rehearsal process and three-week production run, in part because the Gate until very recently was virtually unfunded and didn't pay anybody – I think you got local travel paid as an actor and odd bits of money to pay people. You can't really talk about there being one [single] policy because people brought their own perspectives to it. There are two strands to the translation work, one is that translators have had access to the source language of the text that they were translating. Generally it's not been the case that the Gate has done versions using a literal translation. The other strand is that the translations that the Gate put on right from the start were developed towards staging, were developed towards being seen in performance, very actively, and this is something that all the people that I have talked to about the Gate over the past three years have very strongly emphasized. But there are different ways in which that might take place. We can think of it as a sort of a spectrum and might say that one end of the spectrum is very much about translating the theatrical paradigm – either you transpose the place or you orientate it towards the kind of English or people's imagined idea of an English theatre experience – or you translate much more at the level of language but still looking towards the 'situation of enunciation' as I think Patrice Pavis puts it, but remain, in academic terms more 'faithful', I suppose, to the source text. When the Gate started it was just a venture by an American theatre director called Lou Stein and it was something like a smash and grab raid on plays that he was excited about and this included new writing, adaptations of novels and a very large strand of East European work, and this last strand was the work that excited the critics and everyone involved, not least because the Gate is very, very small and it was even smaller then, it only sat 56 people and Stein was aiming to put on plays with casts of ten or more, Bulgakov's *Crimson Island* and Mayakovsky's *Bedbug*, and produced the kind of clash between the apparent nature of this play and the space in which it was taking place. When Giles Croft took over in 1985 he took on that international policy and made it exclusively the remit of the Gate to bring in lots of different companies to put on plays in translation. He also started an imprint with his publisher brother to publish those translations and try as far as possible to generate new ones. Stephen Daldry took over in 1990 with a strong idea that the programming should focus on British premieres, that it should be the first time that these plays had been seen. Again, I don't think that that actually always happened, every now and then somebody would discover that actually that play had been put on before, but that was the overall aim, the ambition. A significant event in Stephen Daldry's era was director Lawrence Boswell's work on the Spanish Golden Age, which David Johnston was very heavily involved in. This strand developed into a very strong strand of adaptation, with the next Artistic Director, David Farr, who also made it his remit to introduce new European writers to British audiences through two biennales. Looking at the range of work that Mick Gordon, the next artistic director, has done, this is very much about transnational identity, about hybridity, about adaptation, about issues that might not be necessarily single country specific. When Erica Whyman took over in 2001, she was very interested in post-dramatic theatre in Europe and movements towards non-narrative forms, not located in characters, and this does not translate easily to the British stage. She was very brave in her programming at that time, and this is where we go back to notions of performability and translation, the theatre paradigm from one arena to another, but her aim was actually to bring different theatre paradigms into the Gate. When Thea Sharrock took over, she moved to a more director-oriented process in that she was interested to work with particular directors and then encourage them to bring in work that they wanted to do, so, in a way, at that point the Gate slightly lost, to my mind, a clear translation policy as such. Every other artistic director seemed to have a particular ambition about translation. And now, there are two new artistic directors (Carrie Cracknell and Natalie Abrahami) about to announce their programme, so it'll be very exciting to see where it's going to go next, but their work is going to be very much located in 'personality' and really that's where I want to come back to: all of these stages of work, these phases that the Gate has been moving in and out of, and the sort of legacies that have influenced each other have been about the people that have been working there, and it's been very much the case of whether the theatre makers have located the process of translation for the stage in the act of writing (and the relationship between translation and writing is interesting in that sense) or whether it's been taking a translation and then re-imagining it in rehearsal and towards performance.

Q: I'm interested in international work and the kind of commodification of that work. I was interested in some of the metalanguage that you used, import-export, thinking about cultural capital, those kind of words, and I'm thinking that using British playwrights to engage in translation work from other cultures and other languages, who have never lived in that culture, or experienced that culture and don't have first-hand knowledge of the languages . . . is the expectation just about generating a product for an audience?

KM: The thing I would say to you is what you are tackling is in a way the nature of writing. As a playwright you can make absolutely believable the world of another country that you do not have experience of if you are prepared to not just assume, if you can just listen and take it in. There is a very interesting kind of alchemy that happens. I don't exclusively work with playwrights' versions. I myself am a stage translator (I translate from French) but I am tough in my choices. I always want the very best text for my audience and there are times when I really wouldn't pick me as the translator. In fact the majority of the times I wouldn't pick me! If there's a particular play and I think that I am the best person for it, then in theory I could translate that play. We have done a number of contemporary French language plays with Quebec and with France and other Francophone countries, such as African nations, and I am often not the best person, even if I know the language and I am immersed in that culture, not the best person to render those texts on our stage. I do take your point on board but making things believable is the essence/nature of writing. Making something 'real' and believable is about emotional truth, it's about language, it's about all those things put together and in a theatre translation there are so many of these vital elements.

Q: If you see theatre translation as primarily a question of writing, where do you locate the process of cross-cultural communication?

JM: In the route from creative inception to audience engagement there are many different stages and at each one of them there is the risk of commodification, there is the risk of all sorts of good and bad things that come in, and the many choices you have within the confines of your resources, but I can't imagine that one approach would do it. The difficulty is the number because, when Katherine says it all comes down to one, that's the problem for me, there is an immense pressure, for all the different models, all the different opportunities to unravel, to problematize. If you only have one slot where there are many things, that's for me the critical question to come back to.

Q: I am not sure about the role of the literal translator at the Traverse. When you say that you do a version and work with the playwright, do you assume that the playwright speaks English, is it via this other person, are they present, are they absent?

KM: The decision I made was that I would never hide the literal translator in our published texts, and therefore their role is acknowledged in the top of the billing, they are credited. Basically using literal translators isn't a new thing. This has been going on for a long time. But is that translator credited? It throws up a number of ethical questions as David [Johnston] says. The financial one is the next big inroad and actually although we have been making some inroads at the Traverse, progress is slow. It's an endless negotiation, but that's part of the job.

Q: Are the literal translators present or are they absent when you work with the original playwright?

KM: Because English has become a very globalized language, whether we like it or not, many non-British writers have become more able to speak English. It's not a requirement for us: we don't commission a translation of a play because the writer speaks English. If they don't, what we have done in the past is either have an interpreter there, or else what we have done is sometimes to communicate in one of the other languages which they do speak (so, for example, with a Finnish writer we might be able to speak together in French), but you find the best way to do it.

Q: But the literal translator who provides the text is not present?

KM: It depends on the case, sometimes they are, sometimes they aren't. But the literal translation is not performable. We are not asking them for a polished piece of work, we ask them for an information map to the play.

MP: I think that literal translators do play a prominent role in the dissemination of contemporary foreign theatre in Britain, and I would like to start thinking about the 'literal translation' as a first draft of a collaborative project, the beginning of a process, of what you see at the end on stage.

KM: This is how we view it. The other thing is because our PLAYWRIGHTS IN PARTNERSHIP commissions are intended for production and not for readings only, we go through a lot of drafts, and the literal translator will do one of those, (although the reality of that might be two, if they don't show us the first one). There's only one draft from them, but the playwrights might often then go through up to nine further drafts of the translation which is not always usual for translations in Britain.

Q: What happens in terms of copyright?

KM: What you have in our case is that the literal translator will retain rights to their literal translation. Once we as the commissioning UK theatre have done our production, the literal is theirs, and they could take it to another theatre. The English language-speaking playwright retains the rights to their version, as long as it's been approved by the original international writer. If it's not approved by that original playwright, however, nothing can happen. It exists, but it was not going to be performed as it is. Of course, the source language writer has super-rights. It's their play, they wrote it and you don't depart from what they wrote. Furthermore, the English-language rights you have are for a time limited period. You have the English language rights and sometimes it's as specific as 'just for the UK but not for North America', which enables the source language playwrights to get further earnings from an American translation, a Canadian translation, and so forth. But the source language playwright always retains the main rights as it's their play.

Q: You say that the original playwright is involved in the process. But if the playwright is a young, up-and-coming playwright there are economic concerns for him/her too, she or he wants the translation to happen, so they might be thinking 'I'll let them change it'.

JM: But that's not confined to transnational working. It's absolutely the same for domestic playwrights.

KM: We call ourselves 'a writers' theatre' and it's not in our interest to force writers against their wills. If we wanted to determine plays, then maybe we should go and write plays ourselves. What you want to do is to help the writer realize their play, and that's the same with international work: you know you have succeeded when the writer says: 'that's my play. It's not a version of my play, it's my play', and that's what you are aiming for. If the mother swears in a play, what's the equivalent swearword? It's not about changing it, it's about picking the right language, it's about nuances and gradations of strength and ferocity or humour.

Notes

1. www.writernet.co.uk
2. The translation initiative 'Channels' was set up by the National Theatre Studio with the objective of improving the circulation and exchange of new plays in Europe and the world. Playwrights translate foreign plays and meet the foreign playwrights to exchange ideas, and compare practices during the Residency period, when focused workshops for writers and translators take place.

Part IV
Politics, Ethics and Stage Translation

13
The Impotents: Conflictual Significance Imposed on the Creation and Reception of an Arabic-to-Hebrew Translation for the Stage

Yotam Benshalom

The life-cycle of inter-cultural translations, discussed by translation studies and other disciplines, is generally affected by underlying political conflicts between groups and their opposing narratives (see Tymoczko, 2009, pp.171–94; Brownlie, 2007, pp.136–42). Translation studies theorist Mona Baker (2006, p.128) states that any translation between opposing parties in a conflict is constantly affected by political interests and collective ideologies, as translators 'are firmly embedded within specific narratives'. To that she adds (2006, p.1) that translation is constantly used to affect the surrounding conflict in favour of one side or the other, as '[i]n this conflict-ridden and globalized world, translation is central to the ability of all parties to legitimize their version of events'. Baker maintains that conflict-related translations are often used as means of warfare by unjust powers, but that they can be also used as means of just resistance against them. The numerous examples she draws from translations which take place between the western world and the Arab world, often in Middle-Eastern contexts (2006, p.6), exemplify the power and the abundance of translation as a weapon in such disputes. This capacity of conflict-related translations leads her not only to encourage readers to approach translations carefully and critically; it leads her to encourage translators (2006, p.127) to modify their translation products in order to promote their political views, unless they wish to be 'participating in uncritical circulation of a narrative they may well find ethically reprehensible if they stopped to ponder its implications'. Every conflict-crossing translation is ideologically charged, and should be read as such; and since consciously charged translation is already used as a weapon for the sake of deplorable narratives, it would be immoral to refuse to use it as a weapon for the sake of noble ones.

Here I wish to suggest a different opinion. I believe that translators, translation commissioners, critics, spectators and other agents involved in acts of translation in the context of violent conflict should not be encouraged to relate the text in question to one of the conflicting sides and act accordingly. We, as practitioners and observers in the world Baker describes, are well aware of the effects of disputes on translations, and, to some extent, to the effects of translations on disputes. We do not need to be encouraged to treat translations as manifestations of conflicts, as we do so quite willingly. We become skilled in analysing translations, adaptations and other instances of cultural shift, keenly searching them for hidden traces of cultural conflicts. Proud of our critical refinement, we often forget that the conflicts we look for are indeed narratives in their own right, and that we might be reading our favourite conflicts into texts or textual events even when originally they hold no connection to these conflicts. In such cases we hunt for a ghost prey; we plant the traces of conflict in our suspects' pockets. The more prominent the conflicts between groups, the more eager we become to spot evidence for them among processes and products of inter-cultural exchange. This enhances dilemmas we face as translators and addressees alike. Translation of texts intended for theatre performances in particular is especially prone to political impositions upon interpretation, as the multi-staged process of its creation, application and consumption involves many agents and narratives.

This essay aims to exemplify the problematic nature of such imposition by reviewing the processes involved with the Hebrew translation of a single Palestinian play produced on an Israeli stage. It discusses the several phases which this translation has gone through: the initial choice of the play to be produced, the commission of the translator, the translational shifts and deviations from the source text and the reasoning behind them, and finally the reception of the actual translation event and performance by the audience and critics. Being the translator in question, having personally witnessed at least several of these processes, this essay falls somewhere between an objective report and personal reflection upon an experience. The case of *The Impotents* portrayed here should serve as an example in two levels: firstly, for the specific issues involved with the translation of Arabic drama into Hebrew in Israel, and secondly, for the general issues involved with the imposition of conflictual significance upon literary translation between two cultures in a state of violent conflict.

The play and its Hebrew productions

The play العاجزون: اهل العنة, or *The Impotents: The Powerless Folk*, was written in 2001 by the Palestinian poet, playwright, translator and director Riad Masarwy, who lives and works in Nazareth, Israel. It was composed first as a literary work, based on a short story and intended solely for reading, and only later was further adapted by the author and brought to the stage.

The play takes place in a small, humble bar in contemporary Berlin, on a Christmas night which happens to occur during the month of Ramadan. Four Palestinian exiles, regular customers, are gathered together with a Turkish bartender, and try to drink their troubles away in silence. This night, however, one of them is determined to make them all speak. This man, called in the play 'The Suicide', is a theatre director with a taste for classical tragedies; he forces the other characters to expose their painful and humiliating life stories. The stories of all the characters, built gradually on the stage, are themed around disillusionment with everything they had believed in while living in Palestine, their homeland: disillusionment with family, love, religious faith, and, most notably, with Communist-Marxist ideologies which were popular among the Palestinian public in the Israel of the 1970s and the 1980s. The painful disenchantments have rendered the characters sexually and mentally impotent, unable to function and unwilling to talk about their miseries. 'The Suicide' takes extreme measures, including physical violence, in order to make the rest of them speak, and exposes an escalating ladder of humiliations, sins and crimes. 'The Turk', who keeps trying to silence 'The Suicide' to no avail, was imprisoned and raped in Turkey because of his Marxist beliefs; 'The Bearded Man' has lost his wife to a corrupt freedom fighter he used to admire; 'The Handsome Guy' was a political opportunist selling his loyalty and his wife to the highest bidder, and is now searching for a god; and 'The Beauty' is mother to a retarded child, a whore, a drug addict and finally a murderer. Only by violently squeezing these confessions out, resulting in 'The Beauty' cutting her own wrists along the way, can 'The Suicide' gruesomely regain his sexual potency. He is then able to finally admit his own past sin of mother–son incest, an act which is a symbolic desecration of the Palestinian homeland; after which, following the Oedipal precedent, he blinds himself.

The translation of this play into Hebrew was a multi-phased process, spreading over several years. The first person who showed interest in the idea was Nurith Yaari, director of the theatre studies department at Tel Aviv University, who was personally acquainted with the playwright and his work. She wanted *The Impotents* to become a Hebrew student production, directed, designed and performed by theatre students, and addressed to a local and supportive audience of regular subscribers. In 2002 she commissioned it to be translated by myself, one of her students at the time. This production did not take off, but the translated text was not abandoned. Two years later it was adapted for a staged reading in the Cameri Theatre in Tel Aviv, directed by Adva Vardi, a student in the department at the time, and performed by professional actors. Following the success of the reading, the play was included in the annual Acco Festival for alternative Israeli theatre in 2005. The play was directed in Arabic by Masarwy, the original playwright, and performed by a group of young Arabic actors from northern Israel. The final production, eight shows performed within four days, was attended by a mixed group of Arabic- and Hebrew-speaking audiences. My Hebrew translation survived in the form of surtitles, projected above the heads of the spectators during the shows. I accompanied the group during the shows, operating the projection of the Hebrew surtitles.

The Impotents is a unique play in the landscape of Palestinian political drama in general, and in Arabic drama at the Acco Festival in particular, in that it refuses to refer, directly or indirectly, to the Israeli-Palestinian conflict. The clash with Israeli culture is absent from the play, as are questions of war, occupation, discrimination and terrorism. The themes and plot lines are centred around internal Palestinian affairs, ideology and leadership, and are autonomous of any local colonial context; Palestine's crises are outlined against a global background, where the declining cultures of western Europe and traditional Islam deteriorate together into common decay and decadence. From this point of view, there is little interest in the ongoing local fight over a specific piece of land in the Middle East. However, as this essay will show, this conflict still played a major role in the decision-making and interpretative choices of all the residents of the area who were somehow involved in the process of transferring the play into a Hebrew context.

Initial imposition of conflictual significance

Many phases of the translation of *The Impotents* were influenced by the inherent complexity characterizing the majority of contemporary literary translations from Arabic into Hebrew. Tel Aviv University's preliminary choice of the play to be translated and produced was conflict-related and, as such, was made in a quite conscious manner. The usual repertoire of the university's theatre was compiled from ancient and modern canonized

Western drama, so the choice of a local, new and unfamiliar play was far from obvious: the play was chosen not only for its dramatic qualities and performative and pedagogic potential but also for its Palestinian origin and topic. The universities in Israel are in general a spearhead of the Israeli peace camp, and they make many efforts to encourage Jewish-Arabic dialogue. The Tel Aviv University theatre department's decision makers were eager to serve this cause by finding and using a local Arabic play which would be suitable for student production.

If the choice of the play to be translated was political, the ensuing choice of a translator could not possibly be apolitical. The constant conflict between Jews and Arabs in the Middle East has made the Arabic language quite unpopular among Jewish Israeli school students, even though it is an official national language and is used, to some extent, on public television channels, road signs and so on. As shown in research undertaken in the 1990s by Hannah Amit-Kochavi (1999, pp.25–36) and Shimon Balas (1993, p.53), the few working literary translators from Arabic into Hebrew belong traditionally to one of three specific groups, each with its own conflictual ambivalences toward the Arabic language. The first group of translators, most articulate and most influential in the early years of the state of Israel, came from the ranks of Jews who immigrated from Arab countries, especially from Iraq, where Jews were an integral part of cultural activity and humanistic academic life. Many of these Jews have gradually abandoned their heritage of Arabic literacy as part of an effort to fit into the westernized culture of Israel and, for this reason, the number of translators coming from this group dwindles with every passing decade. The second group of translators is centred around the local academic circles. Israeli professors of Arabic language, Middle Eastern Affairs and History of Islam were always active in producing and promoting Hebrew translations from their culture of interest. Here the problem lies with the translation products; these professors are experts in their field, but are generally not accustomed to literary writing. The target texts they produce are often source-oriented, scholarly and linguistically difficult to decipher by regular readers of literature who may not be familiar with, for example, the academic norms of formal Arabic transcription. The last group of literary translators has emerged from a rather unlikely background: the Israeli military. It is a widely known fact that most of the young men and women in Israel go through a period of obligatory military service; some of them become military translators, of texts composed in Arabic, and many of those pursue this craft on leaving the military. The ambivalence of this group towards its work is of special interest. On the one hand, they admire the Arabic language for its beauty and complexity and take pride in their Arabic literacy; but on the other hand, they still tend to see it as the language of the enemy, treating Arabic texts like a cipher which should be decoded with suspicion in mind, constantly looking for hostile intention behind the words.

I personally belong to this third group. From my own experience I can testify that it is still hard for me to translate texts from Arabic without constantly weighing up their relevance to the ongoing conflict in the Middle East, trying to assess if the text should be associated with 'the good guys' or 'the bad guys'. This is true even when the text in question is relatively unrelated to this specific conflict, as was the case with *The Impotents* which deals with internal Palestinian issues against a global background. In the final scene of the play, for example, the protagonist, 'The Suicide', now blinded yet redeemed and self-aware, tells the Turkish bartender he intends to return to his homeland so he can commit his suicide there. As an Israeli military-trained translator, I was startled and even shocked, as I was sure the playwright was talking about nothing less than a suicide bombing. This troubled me, as my personal method of translation often involves attempts to identify with the source author, trying to guess and share his intentions and aims to the best of my ability. Nevertheless, it also made me very careful: I put my emotions aside to the best of my ability and consciously tried to translate the relevant sentences as precisely as possible. I rendered in my translation the Arabic word 'suicide', إنتحار, as the Hebrew word התאבדות – both neutral terms which hold no specific connection to armed hostilities. However, a later discussion with the playwright, held in his home in Nazareth over a traditional meal, proved me wrong. To my great relief, the play written by Masarwy – the author I was aiding and was attempting to identify with – was never meant to imply suicide bombing as a means of personal and political redemption.

Translational decisions[1]

The translation itself was not free of dilemmas and problems. Some of the issues I encountered during the process of transfer were addressed by introducing deliberate changes to the translation, knowingly applying a shift from the source text.

The original play was written for reading, with no intention of production on behalf of the author. Several modifications to the translation were made for the sake of performability, that is, in order to simplify the task of communicating the text to a theatre live audience. Some of the changes were made for the sake of individual actors performing in the Acco Festival production. After noticing that some lines or phrases were being consistently omitted, changed, or mixed up when spoken on stage, I tried to outsmart the actors by modifying my translated subtitles to match. This did not work as well as expected: the following night's performance was the first one in which the actors performed their lines correctly, exactly as they appear in the play, exemplifying some of the inherent unpredictability of such custom-tailored translation for the theatre.

Less temporary, yet just as prominent change, came with another set of performance-related modifications made at an earlier stage, during the initial work done for Tel Aviv University: the addition of linguistic characterization to the dialogue spoken by each character. In the original work all of the speakers use the same homogeneous, relatively high register of literary Arabic, regardless of their personalities and backgrounds. With the blessing of the playwright, I tried to add several gentle touches in order to personalize each and every character and make him, or her, speak in a distinguishable and unique manner; this way I could extend the existing dramatic conflicts further, to the linguistic level.

The intellectual protagonist known as 'The Suicide' is a mature, sensitive and educated man. As an accomplished theatre artist, having directed *Oedipus the King* in his past, he is fond of rhetoric and tends to communicate using long, complex sentences. For this reason I chose to slightly raise his Hebrew register, letting him speak with a somewhat bombastic vocabulary and relatively rich grammatical structures. At the beginning of the play, for example, 'The Suicide' cries in front of his fellow drinkers:

أما أنا فصمتي هو إنتحاري! صمت كثيرا في حياتي... نفذت كل ما طلب مني في النضال صامتا... صفعت وانهزمت, غنيت وبكيت صامتا, وتريدون مني أن أستمر؟ (Masarwy, 2001, p.20)

As for me, my silence is my suicide! I was silent many times in my life . . . I executed silently everything requested from me for the struggle . . . I was defeated and vanquished, I sang and cried silently, and you want me to remain silent?

My Hebrew translation was as follows:

אני שתקתי את עצמי לדעת! רבות שתקתי בחיי... נרתמתי למאבק בשתיקה... ניגפתי והובסתי, שוררתי ובכיתי בשתיקה, ואתם רוצים שאמשיך לשתוק?

I have silenced myself to death! Many were the occasions in my life when I kept silent ... I harnessed myself to the struggle silently ... I was defeated and vanquished, I chanted and cried silently, and you want me to remain silent?

The passionate woman known as 'The Beauty' is a classic femme fatale. She is no less suicidal than 'The Suicide' but, unlike him, she despises words which are not rooted firmly in physical or sensual reality. She is more down-to-earth than any of the delusional men around her; she is impotent morally and emotionally, but not physically. Consequently, I gave her a lower and more passionate register which was closer to vernacular Hebrew. On one occasion, 'The Beauty' tells 'The Suicide':

أنا أكمل حكاياتي متى أردت أنا وليس بأمر منك مثلما قلت لك من قبل. (Masarwy, 2001, p.29)

I will complete my story whenever I want to and not by your command, as I have told you before.

In the Hebrew version, 'The Beauty' says:

כמו שכבר אמרתי לך, אני אגמור את הסיפור שלי מתי שארצה ואתה לא תפקד עלי.

Like I already told you, I will finish my story whenever I like to and you will not boss me around.

'The Bearded Man' is portrayed as a Dickensian hunchback, turned deaf and deranged after listening to too many roars of applause while attending public speeches. I chose to maim him linguistically, making him stutter grotesquely and repeat random words and phrases. The rest of the characters were also subject to a similar, if less noticeable, characterization in translation.

The most significant shift from the original text, though, was not related to dramatic and performative considerations, but to political ones. As mentioned before, the original commissioners of the translation were very keen on staging an Arabic play in an Israeli academic context. There was, though, one problem: this specific play did not have many recognizable 'Arabic' elements in it, as its settings and themes were quite universal. The solution I came up with at the time was simple, if not simplistic: I decided voluntarily to keep the translation as literal as possible, in many occasions rendering word for word,

This modification of register helped maintain an Arabic sound to the Hebrew text, a very noticeable sound which English readers may recognize in some of the early English translations of *Arabian Nights*. It also helped me avoid over-domestication of the foreign play, reminding the actors and the audience that they were dealing with a foreign, Arabic text.

This, however, was also the point at which my excessive awareness of the conflictual context of the translation was beginning to seriously harm my product. It happened because Hebrew and Arabic are not complete strangers: they are both Semitic languages. They have common linguistic ancestors and, to this day, they share common words and grammatical structures. Hebrew became very close to Arabic in lands under Islamic rule during the Middle Ages, but later on, with its revival in the nineteenth and twentieth centuries, has developed in different directions (Snir, 1993, p.25). This means that applying Arabic forms to Hebrew vocabulary does indeed produce valid sentences, but it also gives them an unmistakable mediaeval flavour. I wanted to make my translation sound Arabic; but in doing so I have also made it sound archaic. This archaism in the target text was not equivalent in any way to the original work: *The Impotents* was written in modern day Arabic and deals with modern day characters, even though it consists of fluent, elaborate and somewhat literary language. My changes were there to satisfy the assumed expectations of commissioners, users and audiences; the urge to make the play politically relevant by framing the characters linguistically as explicitly Arab enjoyed dominance, in this case, not only over the ideal of faithful, equivalent translation, but also over the less-praised but no less important ideal of a plausible, communicative target text.

Imposition of conflictual significance by receivers

Israeli audiences and critics often judge theatrical works for their political content, or for the lack thereof, before taking anything else into consideration. The theatre is considered by Israeli viewers to be a potent and serious tool of a political nature, and many public discussions and quarrels are centred around the production of this or that play. On some occasions, judgements upon a production are triggered and passed even when the work itself is 'innocent' of the contents attributed to it by zealous theatre goers. A famous example of this was the reception of Euripides' *The Trojan Women* which was directed in 1983 at Habima, Israel's national theatre, by the German director Holk

Freytag. The Greek tragedy was directed with a contemporary approach in mind, depicting the women as modern refugees guarded by gun-wielding soldiers. Some critics raised an uproar, protesting against the use of Israeli weapons and Israeli military uniforms in the context of war crimes on stage. They related the use of props and costumes in the play to the actions of the Israeli military in Lebanon at the time, actions which were highly controversial for the Israeli public. In the loud and heated debate that followed nobody remembered any more that the uniforms and weapons used on stage were, in fact, not Israeli at all. The uniforms in *The Trojan Women* held no resemblance to those in use by the Israeli Defence Force, and the plastic rifles carried by the soldiers were standard stage props, modelled on the Russian AK-47 (see Levy and Yaari, 1998, pp.99–124). The 'offending' guns and uniforms were wilfully and unconsciously invented by critics who were reading their favourite narrative into the show.

The Impotents did not escape this tendency to impose politics upon theatrical processes during the phase of reception. This inevitable outcome was supported by the surrounding context of the Acco Festival. Acco is a mixed city, inhabited by Israeli Jews and Arabs alike; the festival itself is a framework which emphasizes the co-existence of Arab and Jewish artists, and encourages profound and daring political discussion of the middle-eastern situation. Some theatre researchers even claim that the political awareness of any participating play should be the most important criterion of quality for an Acco production (Levy 2002, pp.70–1). *The Impotents* had two main groups of audience attending the show: local Arabic-speaking Muslims and Christians, comprising approximately 70 per cent of the viewers, and Hebrew-speaking, Jewish theatre enthusiasts, comprising the rest. The reception on the Arabic side, as reflected by local newspaper reviews, was reasonable and straightforward: the critics had treated the play for what it was, and did not 'read' any unrelated political content and issues into the theatrical events. They were concentrating on the professional aspects of the production and also on the play's treatment of internal Palestinian political affairs, mostly criticizing it for 'washing dirty laundry' in public. The case was much more complicated for the Jewish, Hebrew-speaking viewers. *The Impotents* was not an easy target for 'conflict hunters' on the Jewish side, as it was set in Berlin, far away from the disputed Israeli and Palestinian territories, did not contain Jewish characters whatsoever, and was openly declared by its author to be dealing with internal rather than national issues (Shohat, 2005, p.6). Masarwy's play could not easily yield to reductionist views trying to confine it to the public and conceptual narratives traditionally related to the Israeli-Arab conflict. This was probably one of the reasons why most of the Hebrew reviewers concentrated on the production's artistic and performative values, ignoring messages and themes. Initial traces of this puzzlement could still be found in the reaction of the printed press. The best known Israeli theatre critic, Michael Handelzalts (2005, p.4) reviewed the production and wrote: 'I was left to wonder whether it was politics which screwed the sexual and emotional functioning of the characters [. . .] or their sexual and emotional functioning which screwed their politics. But for such a universal issue you do not have to be an Arab.' Like myself at earlier stages, this critic has apparently expected an Arab play to deal with his own concept of an Arabic narrative. Such expectations lead me to believe that when we find ourselves on one side of an extreme political conflict, we become culturally egocentric in an interesting way: we justify the cultural transfer of a work from the other side only when it is directly connected to our conflict with it. We have trouble with the reception of anything not clearly related to that conflict. When we cannot make a direct connection between the text and our own battles, we become confused.

What I found most fascinating was that although the content of the production of *The Impotents* was received in a relatively apolitical manner, my translation was not. Failing to find a hidden political agenda connecting the play or the performance to the Israeli-Palestinian conflict, receivers turned to look for it in the very act and mode of its translation from Arabic into Hebrew, and ended up with surprisingly complicated and awkward interpretations. I personally noticed this during the festival itself. I was operating the projected surtitles of my translation, working with Masarwy and his troupe of energetic actors. While wandering around the festival grounds, set in the scenic ruins of a crusaders' fortress, I overheard two young Hebrew-speaking girls, university students by their appearance, who had just seen the play and were talking among themselves about their experience. What captured my attention was the fact they were interested not only in the show in general, but also in the translation in particular. They tried to understand why the play was produced in Arabic if a Hebrew translation had already existed. One of them suggested that the projected surtitles were intended to create a Brechtian alienation effect; but the other one objected her and claimed that they were a bold political act, a refusal to mask the original Arabic under the hegemony of Hebrew. None of them suggested the obvious: that the production was intended for the Arabic speaking audience, and that the surtitles were there for the sake of

Hebrew speakers who also wanted to come. This dialogue was a unique demonstration of the way theatre spectators can read conflicts of their choice into staged translations: not just into the translated contents, but also into the initial acts of their linguistic mediation.

Imposition of conflictual significance in translation: interpretation and beyond

The acts of imposition outlined here, common to many situations of cultural transfer taking place between two sides of a violent conflict, may suggest a partial explanation of the general unpopularity of simple, straight-forward translations of Arabic plays into Hebrew for the sake of theatrical production. The public nature of dramatic art, being viewed and commented upon by many, tends to transform every Arabic-originated play staged in front of an Israeli Hebrew-speaking audience into a sensational inter-cultural event, its mere existence and frame being more interesting and exciting than its actual content. Of course, the imposition of political significance on Hebrew representations of Arabic culture is not limited to theatre alone. Research in 2007 by a common initiative of Israeli activists and media organizations has shown that more than 50 per cent of the representations of Arab Israeli citizens in local television and radio channels still relate them, in one way or another, to the Israeli-Palestinian conflict (Carmel, 2007, p.5). It is difficult, under such conditions, to approach Arabic works with the aesthetic distance which is necessary, in my opinion, in order to develop a serious, long-lasting tradition of translation – a tradition which would do justice to the source and target texts involved.

Baker's concept, according to which no translation in a situation of conflict should be treated as innocent of hidden political agenda in favour of one of the sides, thus becomes problematic. Its derivative implication – that we, as translators, should use our translations as an extension of conflicts we are part of to promote one narrative and undermine the other – problematizes it even further. This is because narratives are not limited to the story of this or that struggling side alone. Any political conflict is, in itself, a narrative, which is maintained and nourished by many interested parties on both sides, and we ourselves might very well be tempted to join them and actively read it into the texts we encounter. I believe that when we treat translations in that manner, imposing them with political significance with nothing but their context or frame to support it, we lose more than we gain. One obvious loss is the professional credibility of translation, which is discarded by Baker (2006, p.128) as naïve dream at best and collaboration at worst: 'neutrality is an illusion', she writes, 'and thus uncritical fidelity to the source [. . .] has consequences that an informed translator or interpreter may not wish to be party to'. Translators who promote actively one narrative over the other compromise, using Christiane Nord's terminology (2001, p.185), the loyalty to their partners in the source as well as in the target culture. Having betrayed the trust they gained, disloyal translators may run the risk of harming the credit professional translators enjoy (see Chesterman and Wagner, 2001, p.93). A similar risk is shared by doctors who choose to give inferior treatment to injured enemy soldiers. But beyond this ethical, possibly romantic and old-fashioned point of view, lies another issue of a different kind. When we read our own narratives into a text under translation, as is the case with every other imposition of external narratives, we run the risk of reducing the text to the scope of that narrative. As phrased by Brownlie (2007, p.142), '[ideologically] committed approaches [to translation] do not question their presuppositions, so that the particular political stance of each approach is a given.' (See also Hermans, 1999, p.129.) We miss out all content which is not firmly grounded in our pre-existing views regarding the surrounding conflict, and end up learning very little. This, as can be seen in the case of *The Impotents*, holds true for translators as well as for translation commissioners, theatre goers and academic researchers. By imposing a political, conflict-related significance on the cultural mediation of Masarwy's play, the mediators and addressees – myself included – were basing their interpretations and decisions on the context of the work rather than on its content, thus missing some of its potentially unique, unfamiliar messages. If we want to get in touch with 'The Other' we cannot afford the implications of such intentional prejudice in translation, no matter how intense the conflict around us is. Translation can become a formidable weapon, but I believe that it has a mission grander than that. As shown by Baker (2006, p.105), as translators we can never become truly objective, acting like 'detached [. . .] professionals'; but, as the major cultural ambassadors working in the mayhem of a raging battlefield, we have the responsibility to try.

Note

1. For more details about the individual shifts involved in the translation process see Benshalom, 2006, pp.56–69.

References

Amit-Kochavi, H. (1999) *Translations of Arabic Literature into Hebrew: Their Historical and Cultural Background and Their Reception by the Target Culture*, unpublished Ph.D. thesis submitted for the degree Doctor of Philosophy (Tel Aviv: Tel Aviv University)

Baker, M. (2006) *Translation and Conflict: A Narrative Account* (London and New York: Routledge)

Balas, S. (1993) 'Remarks on the translator's textual intervention', in S. Somekh (ed.) *Translation on the Road Margins: Readings in contemporary literary Arabic-to-Hebrew translations* (Tel Aviv: Tel Aviv University Press), pp.53–8. [Hebrew:]

בלס ש. (1993) 'הערות על התערבות המתרגם בטקסט', ש. סומך (עורך) תרגום בצידי הדרך: עיונים בתרגומים מן הספרות הערבית לעברית בימינו (תל אביב: אוניברסיטת ת"א), עמ' 8–53.

Benshalom, Y. (2006) 'From Literary Arabic to Staged Hebrew: On translating *The Impotents* by Riad Masarwy', in *Teatron: An Israeli Quarterly for Contemporary Theatre* (17) (Tel Aviv: General Union of Writers in Israel), pp.56–8. [Hebrew:]

בנשלום י. (2006) 'מערבית ספרותית לעברית בימתית: על תרגום "האימפוטנטים" מאת ריאד מסארווה', תאטרון: רבעון לתאטרון עכשוי, גליון 17 (תל אביב: איגוד כללי של סופרים בישראל), עמ' 8–56.

Brownlie, S. (2007) 'Situating Discourse on Translation and Conflict' in *Social Semiotics*, 17:2, pp.135–50

Carmel, A. (22 October 2007) 'Miriam Toukan IS alone', in *Haaretz*, Galeria Plus Supplement (Tel Aviv), p.5. [Hebrew:]

רמל א. (22/10/2007) 'מרים טוקאן כן לבד', הארץ, מוסף גלריה פלוס (תל אביב)', עמ' 5.

Chesterman, A. and Wagner, E. (2001) *Can Theory Help Translators? A Dialogue between the Ivory Tower and the Wordface* (Manchester: St Jerome)

Handelzalts, M. (4 November 2005) 'Each to her Own: *The Impotents* and *Unstitched* in Tzavta' in *Haaretz*, Gallery Supplement (Tel Aviv), p.4. [Hebrew:]

הנדלזלץ מ. (4/11/2005) 'כל אחת לעצמה: "האימפוטנטים" ו"פרומות" בצוותא', הארץ, מוסף גלריה (תל אביב)' עמ' 4.

Hermans, T. (1999), *Translation in Systems: Descriptive and System-Oriented Approaches Explained* (Manchester: St Jerome)

Levy, S. (2002) 'Have I slept when others were suffering? On theatre and reality here, now', *Teatron: An Israeli Quarterly for Contemporary Theatre* (8) (Tel Aviv: General Union of Writers in Israel), pp.70–1. [Hebrew:]

לוי ש. (2002) 'האם ישנתי כשהאחרים סבלו? על תיאטרון ומציאות כאן, עכשיו', תאטרון: רבעון לתאטרון עכשוי, גליון 8 (תל אביב: איגוד כללי של סופרים בישראל), עמ' 1–70.

Levy, S. and Yaari, N., (1998) 'Theatrical Response to Political Events: The Trojan War on the Israeli Stage during the Lebanon War 1982–1984', *JTD: Journal of Theatre and Drama* (4) (Haifa: Haifa University), pp.99–124

Masarwy, R. (2001) *The Impotents* (Nazareth: Al-Hakim). [Arabic:]

مصاروة ر. (2001) العاجزون (اهل العنة) (الحكيم: الناصرة).

Nord, C. (2001) 'Loyalty Revisited: Bible Translation as a Case in Point', in A. Pym (ed.) The Return to Ethics, Special Issue of *The Translator* (7:2) (Manchester: St Jerome), pp.185–202

Shohat, T. (19 October 2005) 'One Whore, Two Caretakers and Four Impotents' in *Haaretz*, Galeria Supplement (Tel Aviv), p.6. [Hebrew:]

שוחט צ. (19/10/2005) 'זונה אחת, שתי מטפלות וארבעה אימפוטנטים', הארץ, מוסף גלריה (תל אביב), עמ' 6.

Snir, R. (1993) 'Source and Target on the Stitch Line', in S. Somekh (ed.) *Translation on the Road Margins: Readings in contemporary literary Arabic-to-Hebrew translations* (Tel Aviv: Tel Aviv University Press), pp.21–40. [Hebrew:]

שניר ר. (1993) 'מקור ותרגום על קו התפר', ש. סומך (עורך) תרגום בצידי הדרך: עיונים בתרגומים מן הספרות הערבית לעברית בימינו (תל אביב: אוניברסיטת ת"א), עמ' 21–40.

Tymozcko, M. (2009) 'Translating, Ethics and Ideology in a Violent Globalizing World', in E. Bielsa and C. W. Huges (eds.) *Globalization, Political Violence and Translation* (Basingstoke and New York: Palgrave Macmillan), pp.171–94

14
The Politics of Translating Contemporary French Theatre: How 'Linguistic Translation' Becomes 'Stage Translation'

Clare Finburgh

'We are extremely exposed, almost naked. We exist exclusively via language, we can't cling onto anything else.' Actors Christophe Brault and Jean-Paul Dias, who performed in the French première of *Par les routes* by Noëlle Renaude, state (2005, p.17) that the main feature in the play is language. Not character, plot, or issue, but language.

The discrimination in translation studies between 'linguistic' translation and 'stage' translation is today both established and recognized. David Johnston (1996, pp.7, 9, 58; 2004, p.25) distinguishes between scholarly translation concerned with the play at the level of semantic and syntactical units of language; and translation that focuses on dramatic impact and *mise en scène*. Manuela Perteghella (2004, pp.6, 12) differentiates between the translator as *fidus interpres* who produces a reader-oriented translation concerned with the philological exactness of language; and the translator as theatre-maker, who produces a stage-oriented translation concerned with audience reception. Johnston (2004, p.36) states that the communication of character and situation are key to effective translation, since '[a]ctors ... are constantly searching for the emotional truths of their character, forever exploring motivation, hidden agendas and emotional turning points'. And Patrice Pavis (1989, p.27) states that 'The translator is a dramaturg who must first of all effect a *macrotextual* translation, that is, a dramaturgical analysis of the fiction conveyed by the text.' According to these theorists, the translator must consider story, subject or character first, and then find the linguistic means with which to convey them. In the two decades since Pavis advocated this 'holistic' approach, French theatre has evolved considerably. Authors like Michel Azama, Olivier Cadiot, Jean-Luc Lagarce, Philippe Minyana, Noëlle Renaude, Valère Novarina and Michel Vinaver have systematically abolished the theatrical conventions of character and narrative.[1] What is left, is language. Of course, all playwrights construct effects of reality using words. But in the theatre of these writers, reality lies *in* language; reality *is* language. Character and situation become grammatical, running no deeper than their linguistic contours. And yet, whilst displaying great literarity and poetic density, contemporary French theatre also produces the instantaneous effects of a dramatic, and not a poetic text. I should therefore like to ask what happens to the apparently logical and practical opposition between linguistic and stage translation, when the dramatic impact and *mise en scène* to which Johnston and Perteghella rightly accord importance, are produced precisely by the very semantic and syntactical units that they deem to be scholarly and readerly, rather than theatrical.

In this essay, I discuss the theoretical perspectives and the collaborative processes that took place between myself, the playwright Noëlle Renaude, and British writer-actor-director Cassie Werber, explaining how they produced my translation of Renaude's *Par les routes* (*By the Way*). This play was premiered on France's main stage for new writing, Théâtre Ouvert (January 2006), and won the European Theatre Convention prize in 2006. It was then staged by Cassie Werber and her company Chopped Logic at the Pleasance Theatre during the 2008 Edinburgh Fringe. In order to place under scrutiny the binary opposition between stage, and linguistic translation, I interrogate specific terms that are accepted as givens by certain advocates of stage translation, namely 'the stage', 'performability', 'naturalness' and 'breathability'. I argue that these concepts are not neutral; they are economically and politically inflected. The loyalties of the stage translation can lie less with the theatrical specificity of the original, and more with the conventions governing the UK theatre establishment: prioritization of narration, dramatic action, psychological characterization and social thematization. I argue that these terms are therefore ideologically charged in ways that serve to exclude experimental theatre such as that currently produced in France from a culturally imperialist British theatrical system.

The opposition between page-oriented and stage-oriented translation assumes that the latter remains faithful to the dramatic impact, the 'performability' of the original text. However, it strikes me that this claim might be somewhat disingenuous. Considerations of 'playability', 'performability' and 'the stage' dominate discussions of theatre translation (see Brisset, 1996, p.158). They are prioritized by theorists, publishers and artistic directors alike. Translator and theorist David Johnston (2004, p.26) maintains that, 'translators must see themselves to all intents and purposes as writers for the stage'; Michael Earley, representative for Methuen, states the publishers' clear preference for 'stage' versions over academic versions (quoted in Aaltonen: 2000, p.46, note 9); Mick Gordon, Artistic Director of The Gate Theatre (1998–2001) which specializes in foreign playwriting, has reiterated this point.[2] According to these theorists and practitioners, the success of any translation depends on how willingly it adheres to the conventions of 'the stage'.

If translations must accommodate the standards and practices of 'the stage', it is important to understand what this term involves. To my mind, it is not as innocuous as it might appear. Bassnett and Trivedi (1999, p.101) state that there is no sound theoretical basis for a term such as 'performability', since it inevitably varies from culture to culture, period to period, text type to text type. Most significantly, I personally feel, it depends on the economic policies of theatrical systems. The immediate influence and control on translation work for the stage is exercised by this system, which acts as a series of constraints on translators and directors. The theatre and publishing industries determine what is 'performable' in response to what they believe audiences wish to see (see Renaude, 2003; Espasa, 2000, p.56; Dingwaney and Maier, 1995, p.5). In a system such as the UK's, where financial concerns are imperative, 'performability' ultimately means 'marketability'. Even the most heavily subsidized theatre in the UK, the National, receives £16 million in funding, and must raise £50 million from ticket sales. Moreover, in order to secure state funding, all theatre projects must firstly prove their commercial viability. As the French director Jacques Lassalle (1982, p.12) declares, 'It is the period in history, as much as the individual, that determines translation.'[3] The period, which in the UK's case is our current capitalist climate, determines what is 'translatable' and 'performable'. Lefevere (1992, pp.14–15) notes that these economic controls precipitate a form of self-censorship, since translators are only likely to translate texts that conform with publishing and production systems.[4] Michael Cronin (2003, p.4), in a study of translation in the context of global market forces, notes the quietly pernicious nature of this censorship: 'Although more obvious forms of censorship, such as physical assaults on translators or the banning of translated works, tend to attract public attention, less conspicuous and more damaging forms of censorship can go largely unnoticed.' I think, therefore, that theatre and publishing economic ideologies need to be placed at the heart of questions of 'performability' and what lies behind a 'stage translation'.

Since plays that are more or less naturalist, narrative-driven and issue-based are considered to guarantee a British audience, and therefore to be less commercially risky, they tend to be staged. Eminent theatre translator Steeve Gooch (1996, p.15) states:

> It's still the case today that the language of most commercially successful English plays relies on tickling the nerve-ends of our national class-consciousness. A whole range of social responses is invoked by verbal mannerisms, which in turn imply a particular milieu, which in turn enable the audience to 'place' the characters within well-worn social definitions.

Plays attract British audiences if they contain identifiable time, space and social class, all represented in a largely naturalist style. Naturalist theatre is therefore given preference. Phyllis Zatlin (2005, p.12) reiterates: 'Anglo-American realism/naturalism will often get absolute preference over the varieties of more imaginative theatre that come from other cultures in the same way that Hollywood-style movies will dominate over European art films.' The definition of 'performability' is thus determined by commercial imperatives which, in the UK, dictate naturalist form.

The UK has remained resolutely impermeable to the experimental poetic theatre that has developed in France over the past two decades. In 2000, Carole-Anne Upton and Terry Hale wrote (2000, p.1), optimistically, that approximately one-in-eight professional productions reviewed in Britain's national press was a translation. They praised efforts by the Royal Court, Gate, Almeida and National theatres and the RSC (Royal Shakespeare Company), for their support of international writing. However, the one-in-eight to which Upton and Hale refer, contains virtually no new writing. Current productions of foreign plays, for example in French, fall primarily into three categories, all of which exclude contemporary experimental writing. The first consists of classics, for example Racine's seventeenth-century *Phèdre*, staged at the National in 2009. The second contains authors such as Beckett, Ionesco and Genet, who benefit from having been labelled, albeit randomly, members of the recognizable 'Theatre of the Absurd', and from the translation boom that resulted from the post-war hope that cultural exchange might improve geopolitical relations (see Venuti, 1992, p.5). In 2007, Genet's *The Blacks* (1957) was staged at the Theatre Royal Stratford, and Ionesco's *Rhinoceros* (1957) at the Royal Court. The third category contains contemporary plays that adhere to UK codes of naturalism. Most successful of these is Yasmina Reza's *Art* (1994), the only contemporary French play to enjoy commercial success in the West End and on Broadway. Translated into English in 1996 by British playwright and screenwriter Christopher Hampton, *Art* 'tickles the nerve ends' of British audiences, with its psychologically rounded middle-class characters, and its debate on the value of art. Reza's *Life x 3* is also the only contemporary French play to be staged at the National (2000). The only play by Noëlle Renaude to be translated into English is her 1990 *Le Renard du Nord* (*The Northern Fox*, 2000), whereas her plays have been translated into a range of European and non-European languages including classical and regional Arabic, Catalan, Czech, German, Hungarian, Icelandic, Japanese, Polish, Russian, Spanish and Swedish. It is no coincidence that *The Northern Fox* is one of Renaude's only narrative-driven plays.

Many translation theorists involved in the postcolonial field write of the marginalization of foreign theatres because of national, ethnic or cultural elements that are deemed to be incompatible with the British target culture. Anuradha Dingwaney (Dingwaney and Maier, 1995, pp.3–5) speaks of the varying degrees of violence perpetrated against indigenous practices and alien cultures when customs, rites and beliefs, in brief, 'foreignness', are domesticated via processes of familiarization or assimilation; Lawrence Venuti and Douglas Robinson (Venuti, 1998, p.4; Robinson, 1997) write of a scandal in translation arising from the asymmetry between the hegemonic and dominated cultures involved in the translation exchange; Eric Cheyfitz (1991, p.xiv) summarizes the sentiments of these theorists by advocating 'the politics of imagining kinship across the frontiers of race, gender, and class'. To my mind, these critics omit one other fundamental form of marginalization, namely that of aesthetic difference. With reference to the fact that non-European texts have little access to European markets, Susan Bassnett and Harish Trivedi state, in their introduction to *Post-colonial Translation* (1999, p.5), that 'European norms have dominated literary production, and those norms have ensured that only certain kinds of text, those that will not prove alien to the receiving culture, come to be translated'. They do not account for the fact that within Europe itself, amongst neighbours as geographically, linguistically and culturally close as the UK and France, only certain texts come to be translated, namely naturalist texts. An *aesthetic* marginalization is taking place in UK theatre, where plays that are experimental and innovative in terms of form, are either acculturated in order to eliminate their stylistic difference, or else excluded altogether.

The few new French plays translated into English either already adhere to British expectations for naturalist style and content, or else are domesticated in order to conform to British conceptions of 'performability'. There has been a tradition in theatre translation of levelling out linguistic specificities in order to produce naturalist dialogue. For theorist George Wellwarth (1981, p.140), the translator must take into consideration 'speakability . . . the degree of ease with which the words of the translated text can be enunciated'; Jiří Veltruský (1977) recommends the use of short sentences, commonly known words and the rhythms of natural speech, and the avoidance of difficult consonant clusters; Clifford Landers (2001, p.104) insists, 'style, which is by no means unimportant in dramatic translation, sometimes must yield to the reality that actors have to be able to deliver the lines in a convincing and natural manner'. A language that is opaque and resists interpretative mastery must be avoided; a fluency strategy is key. Steeve Gooch, one of the most enlightened British translators, points with regret (1996, p.17) to the systematic processes of dialogue domestication that characterize British translation: 'Unfortunately, the stylistic range of English text-based theatre remains pitifully small . . . so the temptation to "translate" into this diluted range rather than extend into unknown territory is enormous.' Like the concept of 'performability', notions of what seems 'natural' and 'speakable' must also be placed under scrutiny, since they dilute or delete the linguistic opacity and poetic density of the original which, I argue, constitutes the foundation of contemporary French theatre.

Numerous translation studies theorists argue in favour of maintaining not only the content, but also the rhetorical form of contemporary French writing. They stress that the writing styles of Derrida or Lacan, for example, are demonstrative and performative, and must therefore not be translated into descriptive and explanatory language (Graham, 1985, p.15; Venuti, 1992, p.4). Venuti (1992, p.4) describes as 'violent' a situation where the performative play of the linguistic signifier – unidiomiatic construction, polysemy, archaism, neologism, fragmented syntax, discursive heterogeneity, rhythmic regularity – is replaced in translation with linear syntax, univocal meaning, current usage, linguistic consistency, conversational rhythms. He condemns 'the forcible replacement of the linguistic and cultural difference of the foreign text with a text that will be intelligible to the target-language reader' (1993, p.209). Critics have devoted considerable attention to advocating the preservation of language's opacity with respect to the translation of contemporary French philosophy. However, virtually nothing has been written in defence of maintaining the opacity of the linguistic signifier in French theatre, where I feel that the domestication of stylistic difference becomes tantamount to cultural intolerance.

To translate the contemporary French theatre to which I refer into acculturated 'fluid', everyday speech, would betray their very *raison d'être*.[5] One of the driving forces behind the theatres of Renaude and others is a rejection of instrumentalized, devalued everyday language and its reproduction in the predictable and bland speech of naturalist theatre. Novarina, who, in his numerous theoretical writings, crystallizes this generation's attitude towards language, explains that everyday speech suppresses creativity and subjectivity. Language must 'produce. Produce and not represent' (1989, p.29). He, and his peers, seek to unfetter words from utilitarian denotation, and to release them into an elocutionary euphoria; in Antonin Artaud's words (1964, p.172), to 'multiply the possibilities' of language, rather than limiting them. These writers resist what Venuti (1992, p.18) refers to as 'the hegemony of transparent discourse in English-language culture'. Their project therefore has implicitly, though not explicitly ideological motivations. Therefore, to banalize their texts by translating them into common language would remove their principal point of interest and their political potency. As I shall describe, it was therefore my task to distort, pervert and poeticize familiar English, in order to strike the balance between opaque poetry and transparent understandability which characterises today's French theatre.

Contemporary French playwrighting seeks to reinvigorate not only language, but also theatrical form itself. The twenty-five plays that Renaude has written since the 1980s are unified by no particular style or theme. They all, however, share one common feature: an experimentation with, and interrogation of, dramatic form. For this generation of playwrights, linguistic poeticization is inextricably linked to experimentation with *mise en scène* and theatricality. Renaude states (2004, p.148) that her interest is:

> to question theatre, its structure, its function, its modes, its models, to investigate each element that constitutes theatre, or else to locate theatre beyond its limits, to see if it still holds together, to assess to what point one can speak of theatricality, reality, substance; if needs be, to feed theatre to excess or to starve it.

With works like her celebrated *Ma Solange comment t'écrire mon désastre, Alex Roux* (1994–8), she experiments with the limits to which she 'feeds' dialogue 'to excess'. Here, the speech of 2,000 characters, over 350 pages, overflows with volubility as she collapses the apparently clear distinctions between the epic and dramatic, novel and play. In contrast to this 'bulimic' aesthetic, *Madame Ka* and *Fictions d'Hiver* (1998) experiment with the limits to which she can reduce, or 'starve' dialogue, without theatrical communication collapsing. Speech is elliptical, truncated and interrupted. Her latest plays, like *Des Tulipes* (2006) and *Par les routes* (2006) treat questions of theatrical time and space, which she dilates and compresses, testing the ways in which they can be incarnated by the actor's voice and body. With each new work, Renaude poses herself a separate set of formal questions, this becoming the occasion to renew and reinvent both linguistic, and theatrical form.

In turn, each of Renaude's texts poses a new set of questions and risks for directors and actors, who are obliged to expand the conventions of theatrical practice in order to stage her plays. *Par les routes* comprises an uninterrupted succession of verse-like sentences. There is not a single stage direction; not even an indication of who speaks, where, or when. The reader or director must decide how many characters there are; who says what, to whom; what associations to draw between fragments of dialogue. Veteran Renaude director Frédéric Maragnani, who staged the première of *Par les routes*, writes (2004, pp.75, 76): 'the field of contemporary playwriting . . . puts actors, directors and spectators to the test . . . By mistrusting the traditional laws of presentation and production, by challenging them, Noëlle Renaude reinvents (thus obliging others to reinvent) theatre-making.' For many theorists of translation into English, notably Landers, fluency is vital for the actor's delivery. Lars Hamberg (1969, pp.91–2) insists that 'an easy and natural dialogue is of paramount importance in a dramatic translation, otherwise the actors have to struggle with lines which sound unnatural and stilted'; Bassnett (1985) emphasizes the importance of 'breathability', since a translated text must be uttered by the actor without unwanted effort; for Ortrun Zuber (1980, p.93), translators must write for actors. But to which actors do these theorists refer? If they mean those trained in the Stanislavskian method of psychologically incarnating characters, then, again, any translation of these authors is pointless. Renaude (1996, p.59) describes her writing in terms of 'breath' and 'matter' that she 'kneads like dough', indicating the fact that she naturally considers the actor's delivery when writing. But she does not seek to accommodate or indulge the actor. French actors in recent years have had to evolve their practices considerably in order to play texts that incorporate neither character, nor plot, nor theme, but that stage the playful semantic and prosodic possibility of a renewed language which refuses the 'easiness', 'naturalness' and 'breathability' of common speech. Maragnani (2004, pp.81, 79) describes working on Renaude's texts as a 'labour' that physically exhausts actors; it is a foreign language that requires the acquisition of specific vocal and gestural skills in order to be learnt. Veteran Novarinian actor Daniel Znyk (1991, p.16) highlights the great difficulty that Novarina's texts pose to actors: 'Playing Novarina is, for the actor, a terrible challenge that requires courage. You must enjoy putting yourself in the way of danger.' Novarina (2005, p.163) himself describes the breath necessary to utter his texts as that required for 'swimming breaststroke ever more expansively, lengthening breath, movement'. French theatre maintains a respect for the 'irreducibility of the Poet' in the words of Antoine Vitez, great French director and translator (1982, p.7). It is not the text that must modify itself in the rehearsal room in order to accommodate the actors, but the actors who must evolve their practices in order to accommodate the text. The poetic opacity of contemporary French theatre endeavours to reinvigorate both jaded common language, and faded naturalist dramatic practice. Therefore, imposing imperatives of 'naturalness' and 'breathability' would emasculate the counter-cultural potency of this art form.

Jean Jourdheuil (1982, p.35), French director, academic and translator of Heiner Müller and Georg Büchner, writes, 'It is not a question of imitating a work in another language, but of wholeheartedly seizing the gesture(s) that constitute(s) it.' For me, the 'gesture' constituting Renaude's theatre is language. My linguistic approach, which respects the typographic layout, playful polysemy and phonic density of the original, therefore constitutes not just an attempt at a linguistic translation, but at a stage translation. Before I discuss in detail our collaborative processes, I briefly introduce *By the Way*.

Two characters, perhaps men, have just lost their mothers. They console themselves by going on a road trip. And so begins *By the Way*, a 'theatrical road movie', in Renaude's terms. The two men have lost their mothers, and so has everyone they meet along the way. The man in the lay-by where they stop to relieve themselves is returning from his mother-in-law's funeral; the owner of the restaurant where they stop for lunch recounts his mother's suicide; the victims of the fatal road accident they witness are a mother and her four children. And so it goes on. Very literally, the play might pose as a homage to memory of the mother. Mention on a roadside sign of a 'Donkey Festival' on the '24th', might make an oblique reference to the mother of all mothers, the Virgin Mary. The deer with her 'doting mothering eyes' and 'mothering smile' that the characters narrowly miss, might tangentially evoke arguably the most devastating moment in Disney history, namely the death of Bambi's mother. On a more existential level, the play could evoke a general sense of loss and directionlessness. The two characters decide to go to the Atlantic, but end up in the Alps. The modern world, where matriarchal and patriarchal law, grand discourses of truth and belief, have died, or been dismantled, leaves us with a disorientating and bewildering profusion of directions to take and decisions to make. As the two characters mourn the deaths of their mothers, they attempt, both literally and existentially, to map meaning onto a blank and impenetrable world. Conversely, the play could represent a kind of matricide, that liberates the characters from genealogical hierarchy. Like in Jack Kerouac's novel *On the Road*, the French title of which, *Sur la route*, is evoked in Renaude's title *Par les routes*, the open road enables the characters to encounter new directions and values. For Renaude herself, the play presents the ways in which an urbanized world exists via visual signs. Many lines in the play are road signs: signs indicating the highway code ('Careful Children', 'Rumble Strips'); signs indicating directions ('Villain to the left'); billboards selling products or advertising shops ('Strawberries 500m', 'Flowerspace'); petrol signs ('TOTAL'); signs displaying farm names ('The Swallows', 'The Willows') . . . Renaude (2005, p.11) explains how for her, the urbanized world is constructed via a shorthand system of pictograms, words and numbers:

> I had an idea, that of working on a landscape that progresses exclusively via words. I had been on a trip to the South of France . . . and what struck me was that the landscapes were entirely contained within the road signs. These days we can very easily only know places we travel through in terms of the signs that indicate them. The world, for me, is read before it is thought about, it is a world of signs that is deciphered before it presents itself.

France, glimpsed at through a car window, is reduced to a fictional geography, a linguistic landscape. In *Par les routes*, Renaude explores the capacity of words to contain and control realities. Each word is carefully selected to effect a cinematic tracking shot that scans a fictional landscape seen through the car window, zooming in to pick up minute detail on vignette scenes. Here, I provide just some of the possible readings of this 'road poem' that, like all poems, permits multiple and varied interpretative versions.

However, to read Renaude's theatre purely in terms of narrative or theme, often leads to dead ends and circuitous routes. In *Par les routes*, the plot, like in the majority of Renaude's plays, is a *faux* plot. Like the characters in *Par les routes*, who stray off the motorway, Renaude herself deviates from an Aristotelian linear trajectory, filling the dramatic space with a multiplicity of dissipated details with no central core. Coherence in Renaude's theatre is neither narrative, nor thematic, but linguistic. The homonym in French of *mer*, meaning 'sea', is *mère*, meaning 'mother'. The two characters leave Paris on a geographical journey to find *la mer* and on an existential journey to refind *la mère*. But essentially, they travel on a linguistic and phonetic journey through language. At the grocer's where the pair stops, the arthritic shopkeeper cries, 'Tous ces maux', literally meaning 'all this pain'. The attentive listener will note that *maux* is a homophone of *mots*, meaning 'words'. She therefore also cries, 'All these words', which are the essential substance of the play. I felt that my priority in translating this play, and this author, was to devote my attention to *language*, rather than to concerns with character, situation or location.

The play traces a precise route across France, that can be plotted on a map.[6] The forests, rivers and tourist attractions feature in the text in the strict order in which they appear geographically if one drives south-west out of Paris. Since an audience from outside Paris might not recognize the geographical locations any more or less than a British audience, I chose to maintain the original names instead of naturalizing settings and cultural references. I felt that the potential difficulties that actors might incur when pronouncing the French names, and the absence of geographical and cultural reference points for the British audience, could advantageously augment the two friends' sense of disorientation.

My translation process involved identifying the linguistic and stylistic semantic and acoustic features with which Renaude played in the original, evaluating their dramatic effects, and attempting to create parallel effects with English words and sounds. Renaude – the playwright – and I – an academic specializing in French theatre and language – worked in intensive collaboration on extracts of the text, that we scrutinized word by word, syllable by syllable, sound by sound, in order to identify the formal linguistic features with which she was experimenting. These extracts then provided me with a template that I then applied to the rest of the text. This extract contains just some examples of Renaude's wordplay:

> Quand je serai morte et ça ne saurait tarder Caroline tu ne pourras plus me contredire tu regretteras de m'avoir dans ces circonstances épouvantables contredckkt : madame de avale tout droit sa menthe
>
> : mademoiselle de crisse

Je m'étrangle : madame de vacille

Et tu ris

Je m'étouffe elle rit

Je tente de nous sortir de nos ennuis la honte bientôt va me tuer et tu t'en balances sale petite conne égoïste

Madame de tend ses vieilles mains : je n'ai pour tout bien que cette voiture allemande et ma fille et j'ai besoin de ma voiture allemande pour notre petit job (2005, p.36).

***

When I am dead and this will not be long in coming Caroline you will no longer be able to contradict me you will be sorry for having in these appalling circumstances contradckkt : Lady   of   swallows straight down her mint

: Lady   of   the Younger squeaks

Im choking : Lady   of   reels

And you laugh

Im suffocating she laughs

I am trying to extricate us from a difficult situation the shame soon will kill me and you dont give a shit you dirty selfish little fuck

Lady   of   holds out her old hands: all my worldly possessions amount to this car from Germany and my daughter and I need my car from Germany for our little business

The two travellers park up between rows of lorries along an A road. They run into a sex worker, Mademoiselle de, pimped by her mother, Madame de. Firstly, 'de' in a French surname denotes aristocratic lineage. Renaude deletes the main part of the surname ('d'Estaing', for example), in part preventing psychological identification with the characters, and also obliging the actor to make a glottal pause after 'de', in turn generating a rhythm that audibly evokes Madame de's choking. Since the adverbial indication of title, 'de', meaning 'of', does not carry the same denotation of nobility in English, I had to find another equivalent that assured guttural pronunciation. I chose Lady   of   , and Lady   of   the Younger, the typographical spaces denoting the actor's necessity to mark a pause. Secondly, the scene is driven by abrupt shifts in tone between the haughty register denotative of the family's noble status, and the obscenities more befitting of an insalubrious lorry park. The inversion of *soon* and *will* in 'the shame soon will kill me' affords the dialogue a preciousness that contrasts starkly with Lady   of   's ensuing string of obscenities. In addition, the sonority of the long vowel sounds and sibilants in 'shame soon', and the lyrical rhyming of 'will kill', are thrown into relief by the clipped vowels contained in the swear words. Thirdly, I had to find English equivalents for Renaude's blocks of sound that characters almost spit out as a seemingly unmediated expression of their physical sensations. For example, in 'Lady   of   swallows straight down her mint', the phrase 'straight down' interrupts the usual syntactical order, as if the sentence itself were choking.[7] Finally, like Renaude, I omitted almost all punctuation. Renaude resists using punctuation so that actors must reflect accurately on where she positions pauses. Palpable sonority is central to Renaude's writing. The two actors who performed in the première, explain: 'It's not so much what the play says that's important, as how it's said, the sound it makes . . . In Noëlle's work, we are speakers, everything is in the language; it's like speaking a musical score' (Brault and Dias, 2006, p.17). Renaude chooses each syllable in order to draw melodic lines and rhythmical patterns. By not including punctuation, actors can notice, for example, that 'Im choking' should be contracted, since Lady   of   struggles to articulate her words, whilst 'I am trying to extricate' should be pronounced in full, in order to stress the line's literary ostentation. Here, I indicate just some of Renaude's linguistic effects, that I worked into the English translation. I found that the effects were at times more exaggerated in the translation than in the original. Vitez (1982, p.9), remarking on his own translation of practice, says that, when faced with linguistic irregularities in the original, the translator has the choice either of 'under-translating' and omitting an element, or else 'over-translating', and hyperbolizing it. This advice often informed my choices as translator.

In our workshops together, Renaude frequently described her language as 'faux familier', and explained her processes of 'syntactical disordering'. This short extract illustrates how she takes familiar expressions – she rarely resorts to unusual words or to lyricism – and changes certain elements in order to revitalize them semantically or rhythmically. Together, we selected stylized cells of sound and sense to create a 'falsely familiar' form of poeticized English. We chose to spend our week together in Tours, where *Par les routes* was on tour (November 2006). Each night, after working on the translation, we watched the play and listened to it; listened to the audience's reactions; informally spoke to audience members to gain insights into whether they found the language accessible, opaque, poetic or everyday; interviewed the director and actors. I then pieced together information, impressions and sensations, in order to ascertain what effects the play created in dramatic terms.[8] When back in the UK, I attended rehearsals of the play staged by director Cassie Werber and actors Stavros Demtraki and Kevin O'Loughlin, in order to make adjustments and to gauge whether I had found an equivalent for Renaude's sound poetry of the everyday.[9]

Director Jacques Lassalle (1982, 12) states, 'We must preserve these lumps and bumps, these moments in the foreign text that become so many signs of what cannot reach the other shore. We must conserve what is uncompromising in the text, like reefs that emerge as we draw towards the port.' I have explained my approach towards translating a work of contemporary French theatre. I have resisted complying with theatrical conventions such as naturalism, character coherence, plot linearity or 'breathability'. Instead, I have sought to produce in English Renaude's effects of syntactical inversion, guttural ellipsis, tonal aberration, typographical innovation, semantic polyvalency and dramatic experimentation. In conclusion, I should like to propose that my adherence to the linguistic specificity of the French text rather than to the dramatic conventions of UK theatre, might constitute a political act.

In a recent article, Venuti (2008, p.18) discusses ways in which translation 'might constitute a cultural means of resistance that challenges multinational capitalism and the political institutions to which the global economy is allied'. He illustrates with instances where translators 'short-circuit or jam' the politics of the original text by altering its ideological partiality (2008, p.21). In my case, translation becomes political intervention not by tampering with the original, but rather, by respecting the original's status as Other. Venuti states elsewhere (1992, p.5) that a fluency strategy which effaces the text's linguistic difference in order to render it more intelligible and familiar to the target audience, provides '[the spectator] with the narcissistic experience of recognizing his or her own culture in a cultural other'. The insistence on translating texts that conform with a domestic audience's expectations accords this audience the smug impression that they can understand anything from anywhere. A translation approach that respects the linguistic or cultural specificity of the original text therefore becomes a political necessity. The translator becomes what Carole-Anne Upton and Terry Hale (2000, p.2) terms 'cultural advocate'. As Vitez (1982, p.6) exclaims, 'We are called before a world tribunal to translate; it is almost a political and moral obligation'. Instead of indulging the audience by reproducing recognizable and acceptable forms, the translator has the political responsibility to 'give the public's taste a slap' (Vitez, 1982, p.8).

Translation that obliterates the linguistic or cultural specificity of the original is not only politically imperialist, but it also advances cultural parochialism, since the exposure of UK audiences to other cultures diminishes (Venuti, 1992, p.6). Moreover, the encounter with different cultural contents and linguistic or dramatic styles is not just an altruistic gesture towards the foreign other. As Sirkku Aaltonen states, all translation is to some extent a form of egotism, motivated by self-interest; the foreign is always communicated in response to the needs of the target culture (2000, p.1). Howard Barker (1998, p.79) writes, 'the emphasis on clarity and realism has effectively abolished poetry from the [British] stage'. The domination of naturalism and issues-based theatre, dictated by the blind market forces that drive the British theatrical system, stifles creativity. Liberal economics have constituted a move in theatre not towards complexity and variety, but towards homogeneity and stagnation. As Venuti insists (2008, p.32), the translator has a social responsibility not to acquiesce before the *status quo*, but rather to submit it to searching critique. Exposure to other aesthetic forms therefore becomes beneficial to the target culture, that can renew and reinvigorate itself. In this way, powerful cultures depend upon less dominant cultures (Perteghella, 2004, p.6; Venuti, 1992, p.9). Danièle Sallenave, French translator of Italo Calvino and Luigi Pirandello, writes (1982, p.23), 'For me, translation involves continuing to learn a foreign language and at the same time, continuing to learn my own language. It involves making it say things that it has never said.'[10] If French theatre texts do not enter conveniently into UK frames of reference, this surely reveals the limits of the frames themselves (see Nikos Papastergiadis, 1990, p.100). Foreign languages, styles and genres must be allowed to interrogate, even to radically disrupt the target culture. Therefore, even if contemporary French theatre is not currently in any way recognized by the British theatre establishment, I still feel it is vital to translate it, in order to enable encounters with its exuberant linguistic brilliance, and exhilarating theatrical experimentation.

I end by dedicating these words by Steeve Gooch (1996, p.14) to the people with whom I collaborated: playwright Noëlle Renaude, French actor Christophe Brault, and UK director Cassie Werber:

> Translating plays can only be an act of love. For me it invariably has to do with discovering in the original play some new and slightly exotic quality quite outside the more familiar ground of home-grown plays, which I feel the home audience should know about: like a love affair with a fascinating foreigner whom you feel compelled to introduce to your family.

Without their generous time and invaluable expertise, my 'linguistic translation' would never have become 'stage oriented'.

Notes

1. See Ryngaert and Sermon (2006) for an eloquent discussion of the deconstruction of character in contemporary French theatre.
2. He made this comment in a presentation at the *Staging Translated Plays* conference, University of East Anglia, June 2007.
3. This, and all subsequent translations (except Pannwitz in Benjamin), are my own.
4. Academics are maligned for producing scholarly translations that are theatrically unusable. I believe that the university establishment nonetheless plays a vital role in the translation and dissemination of foreign texts, that are increasingly deemed by the commercial sectors of translation, publication and production, as culturally inappropriate and financially non-viable. The internet has revolutionized the world of contemporary experimental poetry, which for decades was ignored by conventional publishers. It could also be enlisted to save contemporary avant-garde theatre.
5. I do not dismiss translation into everyday English categorically. Douglas Robinson (1997, p.9) argues convincingly that some texts which stubbornly maintain their foreign features can appear either esoteric, or childlike, thereby doing the original a disservice.
6. Céline Hersant has plotted the route's coordinates (in Renaude, 2006, p.104).
7. Here, I have been informed in particular by the theatre of Howard Barker. The way in which he describes his dialogue demonstrates how akin he is to contemporary French playwrights:

   > It is a speech as contrived as poetry, dislocated, sometimes lyrical, often coarse, whose density and internal contradictions both evoke and confuse. In a culture in which language has lost its public status in favour of image and selling, this flood of verbal sound overwhelms the listener, who must

be content with a partial understanding but whose attention is fixed by the sensuality and substance of speech in the mouth of the trained actor. (1998, p.81).

8. I also had a DVD recording of the play, which I could later watch, to check the semantic and rhythmical nuances of the text.
9. Cassie Werber has written and directed three plays with her company Chopped Logic: *Paramour* (2005, Pleasance, Edinburgh), *The Runaround* (2006, Camden People's Theatre) and *Double Negative* (2007, Oval House). I, and Renaude will be present at rehearsals, to assist with necessary modifications to the translation.
10. She echoes Rudolf Pannwitz quoted by Walter Benjamin (1968, p.81):

> The basic error of the translator is that he preserves the state in which his own language happens to be instead of allowing his language to be powerfully affected by the foreign tongue. [. . .] He must widen and deepen his language by means of the foreign language.

References

Aaltonen, S. (2000) *Time-Sharing on Stage: Drama Translation in Theatre and Society* (Clevedon: Multilingual Matters)

Artaud, A. (1964) *Le Théâtre et son double* (Paris: Gallimard)

Barker, H. (1998) *Arguments for a Theatre* (Manchester: Manchester University Press)

Bassnett, S. (1985) 'Ways Through the Labyrinth: Strategies and Methods for Translating Theatre Texts', in T. Hermans (ed.) *The Manipulation of Literature* (London: Croom Helm), pp.87–102

Bassnett, S. (1991) 'Translating for the theatre: the case against performability', *TTR: Traduction, Terminologie, Rédaction* (4: 1), pp.99–111

Bassnett, S. (1999) with Trivedi, H. *Post-colonial Translation* (London, NY: Routledge)

Benjamin, W. (1968) 'The Task of the Translator', *Illuminations*, Harry Zohn (trans.) (London: Fontana), pp.69–82

Brault, C. and J.-P. Dias (March 2006) Interview with Pascale Gateau, *Théâtre Ouvert Le Journal* (15) pp.17–18

Brisset, A. (1996) *A Sociocritique of Translation: Theatre and Alterity in Quebec, 1968–1988*, R. Gill and R. Gannon (trans.) (Toronto: Toronto University Press)

Cheyfitz, E. (1991) *The Poetics of Imperialism: Translation and Colonization from The Tempest to Tarzan* (New York: Oxford University Press)

Coelsch-Foisner, S. and Klein, H. (2004) *Drama Translation and Theatre Practice* (Frankfurt: Peter Lang)

Cronin, M. (2003) *Translation and Globalization* (London, New York: Routledge)

Dingwaney, A. and Maier, C. (eds.) (1995) *Between Languages and Cultures: Translation and Cross-Cultural Texts* (Pittsburgh and London: University of Pittsburgh Press)

Espasa, E. (2000) 'Performability in Translation: Speakability? Playability? Or just Saleability?', in Upton (2000), pp.49–62

Gooch, S. (1996) 'Fatal Attraction' in D. Johnston (ed.) *Stages of Translation* (Bath: Absolute Classics), pp.13–21

Graham, J. F. (ed.) (1895) *Difference in Translation* (Ithaca, London: Cornell University Press)

Hale, T. and Upton, C.-A. 'Introduction', in: C.-A. Upton (ed.) (2000) *Moving Target: Theatre Translation and Cultural Relocation* (Manchester: St Jerome)

Hamberg, L. (1969) 'Some Practical Considerations concerning Dramatic Translation', *Babel*, (15: 2), pp.91–100

Johnston, D. (ed.) (1996) *Stages of Translation* (Bath: Absolute Classics)

Johnston, D. (2004) 'Securing the Performability of the Play in Translation', in S. Coelsch-Foisner and H. Klein, pp.25–38

Jourdheuil, J. (March 1982) 'De quoi parlions-nous?' *Théâtre/Public* (44), p.35

Landers, C. (2001) *Literary Translation. A Practical Guide* (Clevedon: Multilingual Matters)

Lassalle, J. (March 1982) 'Du bon usage de la perte', *Théâtre/Public* (44), pp.11–13

Lefevere, A. (1992) *Translation, Rewriting and the Manipulation of Literary Fame* (London: Routledge)

Maragnani, F. (Spring 2004) 'Le projet de *Ma Solange*', *Frictions* (8), pp.75–91

Novarina, V. (1989) *Le Théâtre des paroles* (Paris: P.O.L.)

Novarina, V. (2005) 'Attraction *chantier 118*', in Nicolas Tremblay (ed.) *La Bouche théâtrale: Études de l'œuvre de Valère Novarina* (Montreal: XYZ), pp.159–72

Papastergiadis, N. (Summer 1990) 'Ashis Nandy: Dialogue and the Diasopra – A conversation', *Third Text* (11), pp.99–108

Pavis, P. (1989) 'Problems of translation for the stage: interculturalism and postmodern theatre', in H. Scolnicov and P. Holland (eds.) *The Play Out of Context* (Cambridge: Cambridge University Press), pp.25–44

Perteghella, M. (2004) 'A Descriptive-Anthropological Model of Theatre Translation', in S. Coelsch-Foisner and H. Klein, pp.1–24

Renaude, N. (Decembre 2003–May 2004) 'Une opiniâtreté paradoxale qui engendre un "gain d'intérêt"', *Actes du théâtre*, p.18

Renaude, N. (2004) 'Paroles d'auteurs', in P. Minyana, *Prologue, Entente cordiale* et *Anne-Marie* (Paris: Théâtre Ouvert), pp.147–9

Renaude, N. (2005) *Par les routes* (Paris: Théâtre Ouvert)

Robinson, D. (1997) *Translation and Empire: Postcolonial Theories Explained* (Manchester: St Jerome Publishing)

Ryngaert, J.-P. and J. Sermon (2006) *Le Personnage théâtral contemporain: décomposition, recomposition* (Paris: Théâtrales)

Sallenave, D. (March 1982) 'La Mise en mouvement d'une langue', *Théâtre/Public* (44), pp.20–3

Veltruský, J. (1977) *Drama as Literature* (Lisse: Peter de Ridder Press)

Venuti, L. (ed.) (1992) *Rethinking Translation: Discourse, Subjectivity, Ideology* (London, New York: Routledge)

Venuti, L. (1993) 'Translation as Cultural Politics: Regimes of Domestication in English', *Textual Practice* (7.2), pp.208–23

Venuti, L. (1998) *The Scandals of Translation* (London: Routledge)
Venuti, L. (2008) 'Translation, Simulacra, Resistance', *Translation Studies* (1: 1), pp.18–33
Vitez, A. (March 1982) 'Le Devoir de traduire', *Théâtre/Public* (44), pp.6–9
Wellwarth, G. (1981) 'Special Considerations in Drama Translation', in M. Gaddis Rose (ed.) *Translation Spectrum: Essays in Theory and Practice* (Albany: State University of New York Press), pp.140–6
Zatlin, P. (2005) *Theatrical Translation and Film Adaptation: A Practitioner's View* (Clevedon: Multilingual Matters)
Znyk, D. (September 1991) 'L'orgueilleuse modestie de l'acteur', *Théâtre/Public* (101–2), p.16
Zuber, O. (1980) *The Languages of Theatre. Problems in the Translation and Transposition of Drama* (Oxford, New York: Pergamon Press)

15
Translating Zapolska: Research through Practice

Teresa Murjas

Introduction

I am a translator and director of Polish plays, born in the UK. My academic background is in Theatre Studies and I work as a lecturer, teaching and researching within this subject area. My ability to speak Polish is rooted in my background and was never developed in an academic context when I was a student. As such, there is a strong link within my research – reflected in an interest in the ethics of translation – between a language I learnt in specific post-war diasporic community contexts (wherein theatrical practice was supremely important as a means of exploring concepts of national and cultural identity) and the skills I have learnt as a student and academic working within a defined, professionalized field.

During my eight years as a theatre lecturer at the University of Reading, I have tentatively started to work on what I hope will evolve into a long-term research project. Its main focus is fin-de-siècle Polish theatre and dramatic literature. The project is therefore a work in progress, yielding a variety of research outcomes. These include published translations and historical research, as well as research performances which I direct and stage at a variety of venues in the UK and in Poland. The outcomes clearly indicate the fact that the research is inter-disciplinary and draws on discourses and methodologies developed within the areas of Theatre Studies, Translation Studies and Slavonic and East/Central European Studies. Its main characteristic, however, is a strong focus on embodied, dynamic and collective translation practice as a means of academic investigation and as a spur to instigating interaction between particular communities and audience groups. So far, the plays of Polish writer and actor Gabriela Zapolska have provided anchorage for these activities, and two books have evolved from my translation and direction of four of her plays: *The Morality of Mrs. Dulska* (2007) and *Zapolska's Women: Three Plays* (2009). My forthcoming book, *Invisible Country: Four fin-de-siècle Polish plays* (2010) will consider works written by her contemporaries; Tadeusz Rittner, Włodzimierz Perzyński, Stanisław Przybyszewski and Jan August Kisielewski. At the time of writing this chapter, I am preparing a production, in a new English translation, of Włodzimierz Perzyński's 1906 play *Ashanti Girl*. What binds all these practitioners together is their interest in naturalism as a theatrical form.

Research contexts

My developing research intervenes in a prevalent culture of interest – within professional UK theatres and university drama departments – in the ground-breaking practices of mid-late twentieth/twenty first-century Polish theatre makers who took a radical deconstructive approach to dramatic literature or devised their own work. Key examples are Jerzy Grotowski, Tadeusz Kantor and the company Gardzienice. These practitioners have frequently been described – and have described themselves – as avant-garde. However, explanations regarding how and why, in their own context, this may have been the case, have often taken, in my opinion, insufficient account of the extremely ideologically problematic concept of 'the Polish mainstream', the role of naturalism and realism as theatrical forms within it, and the impact of literary/theatrical censorship prior to the Second, and indeed the First, World War. Study of these practitioners' work has occasionally been conducted without extensive engagement with translations of their literary Polish source texts – sometimes due to a lack of availability – and often focuses specifically on more frequently translated, documented instances of performance or systems of rehearsal in which the role of language has been de-stabilized to allow the emergence or foregrounding of other modes of communication. The practitioner's work is then theorized accordingly. Such an approach may in certain circumstances be considered methodologically problematic, though I am not suggesting there is anything politically sinister about it. Nor am I attempting to overvalue the role of the play text within the field of Theatre Studies or devalue the role of devised theatre and performance in Poland, or indeed its study. The British Grotowski Project at the University of Kent, for example, is highly significant in terms of its impact on theatre scholarship. However, I would argue that historically conditioned factors, which to some degree continue to affect accessibility and which have affected Polish academics' own ability to research, publish and disseminate their work both within their own country and abroad, do still exist as a legacy. It is also indisputable that this legacy is progressively being challenged.

In a broader scholarly context, this relative paucity of resources makes it extremely challenging to teach an undergraduate module on Polish theatre from a historiographic perspective within a drama department, due to the limited availability of corresponding translated primary, and secondary, resources produced within the field of Theatre Studies that draw on shared critical contexts and discourses. Our recently established and highly successful course at Reading, for example, entitled Polish Film and Theatre, draws on world-leading critical material generated by UK academics relating to the work of Grotowski and Gardzienice. However, we still struggle *not* to reflect a 'historical narrative' that is built on the *lack* of translated primary texts and of critical material that fails to take account of dramatic literature *in performance*. In effect, this *lack* must be expressed reflexively as part of the teaching process. It is undeniable that excellent translations of plays by Witkiewicz, Mrożek and even Fredro, for example, exist, many originating in the US or that superb scholarship continues to emerge within the field of Slavonic/East European studies. However, the subject-specific, overall limitations I have described do remain in place, conditioning how UK students and academics who cannot speak Polish conduct research into Polish theatre within drama departments. They are felt most keenly with regard to late nineteenth/early twentieth-century naturalist/early modernist play texts falling within my own area of research. The teaching of theatrical naturalism, crucially important to most undergraduate university theatre courses on European theatre in the UK, rarely includes Polish examples and is, in any case, frequently 'silent' on the note that reading a text in translation is an important 'issue' (this being seen as the preserve of requisite language departments). Additionally, broader socio-political/geographical concerns relating to Poland's current and historical place in Europe continue to determine the ideologies of academic curricula, impacting on how naturalism is conceptualized as a theatrical form. What I perceive as an unbalanced situation perhaps arises from persistent insufficient negotiation between so-called source and target culture via translation, curious given Poland's now not so recent accession to the EU and the emphasis on both 'new' and 'old' writing in British theatre. Whatever culture of exchange does exist is weakened by the extreme lack of confidence demonstrated by UK editors, many of whom consider publication of play texts only 'on the back of' professional productions; a catch-22 situation, since theatre producers/directors are just as reluctant to take a financial gamble with plays that may be viewed – to put it very crudely – as alien, exotic, marginal or irrelevant by critics. There is no vibrant performance tradition of translated Polish dramatic literature on the British stage – no such dynamic expression of an engagement between these countries' literary histories in this forum. At the time of writing this chapter, Tadeusz Słobodzianek's excellent play *Our Class* (2009) is being staged – in an English version by playwright Ryan Craig – in a groundbreaking, highly-acclaimed production at London's National Theatre. It may be that this signifies a cultural shift, since the preference, historically, has been for hosting visiting Polish companies as part of International Festivals. When it comes to a figure like Gabriela Zapolska – a canonical and extremely popular female playwright in her own country, yet never professionally performed in the UK – the situation attains further complexity.

Zapolska and her contexts

Maria Gabriela Stefania Korwin-Piotrowska (1857–1921) was born in Polish Galicia. She grew up in Podhajce, near Łuck, in the area then called Wołyń. This was once part of Eastern Poland and is now Volhynia, part of Ukraine. She took the stage name Gabriela Zapolska and also wrote under the pseudonym Józef Maskoff. She was a commercially successful playwright and an actress, working for a short time at Antoine's Théâtre Libre and Lugné Poë's Théâtre de l'Oeuvre in Paris. She also worked as a journalist and was a prolific prose writer. Her reputation as a glamorous social outcast stemmed initially from her divorce from a military man and an affair with a married theatre director. The child they had together subsequently died. Zapolska's attempt to sue a national newspaper provoked further scandal, as did her willingness to discuss identity politics and economics from a fundamentally left-wing perspective. Dubbed the 'immoral actress', she and her work became a site for emotive debates about the role of women in Polish society. Zapolska's persona and voicing of alternative and oppositional discourses on national identity, gender, religion, ethnicity and sexuality significantly raised the profile of her work.

During the playwright's lifetime Poland was partitioned by Russia, Austria (later Austro-Hungary) and Prussia and did not exist on the map until after World War I; as expressed in Alfred Jarry's play *Ubu Roi*, it was, literally, a kind of 'nowhere'. This socio-geographic fragmentation had profound consequences in all private and public spheres. Events in one partition did impact on the functioning of the other two, especially given the repeated insurrections that characterized the Russian partition in particular throughout the nineteenth century. However, the nature of government and law enforcement within each partition varied immensely. Since Polish historiography has its nationalist, romantic and Marxist strands, any attempt to recount and encompass the events of 123 years of nineteenth/early twentieth-century Polish cartographic 'invisibility' presents a considerable challenge and post-colonial theorists and revisionists may continue to provide important strategies for re-generating and re-defining the historiography of this period (*particularly* taking account of Poland's cultural diversity). The key for the historiographer and translator must be to note her or his ideological position in relation to events during those 123 years. The extent, to which an outline of seventeenth/eighteenth century geographical boundaries is 'permitted' to hover in the consciousness over maps drawn up between 1795 and 1918 from which it was eradicated, and the significance assigned to this 'presence', is key.

Polish scholars, predominantly discussing the work of canonical male writers, have frequently identified the unifying characteristic of native Polish literature written during the period following the brutally suppressed January Uprising of 1863, as tension between neo-Romantic nationalism and compromise. It has been noted that they were forced to communicate via metaphor, allusion and subtext. Norman Davies suggests that the public expression of nineteenth century Polish politics was so harshly repressed by police and censors that its only outlets were the metaphors of poets and the allegories of novelists. These ultimately came to constitute an ideologically potent, codified and symbolic means of communication. He concludes that such literary texts were crucial to the internal politics of partitioned Poland but are ultimately untranslatable on account of what he perceives as a sort of semantic insularity (Davies, 1986). This factor is identified as an almost insurmountable challenge for theatre translators and directors of Polish texts outside Eastern Europe, including those written after World War II.

Though to some degree I sympathize with this assessment, acknowledging the often irretrievable losses incurred via processes of censorship, I also remain unconvinced that the politics of translation and publication of Polish (or any) literature are thus reducible, most particularly in relation to drama and live performance. The material and ideological systems within which translators and non-Polish speaking theatre practitioners have had to function have surely affected the dissemination and reception of this work.

What we know of Zapolska's theatrical practice acts to problematize such arguments. This applies equally to the work of other naturalist playwrights of this period, which is rarely the subject of contemporary critical investigation in Poland, let alone abroad. Zapolska's formal and stylistic pre-occupations, in a corpus of approximately forty plays, stem directly and vividly from the political and social circumstances in which she functioned and which she, in addition, sought to satirize. She has been referred to as both the 'Polish Zola' and the 'Polish Molière' – titles reflective of formal tensions in her work. However, her gravitation towards naturalism indicates a strong interest in the possibilities of theatrical form as mimesis; as an act of resemblance or a representation of society functioning both within, and parallel to, the contemporary – the 'everyday'. Thus, she attempts a dramatic distillation of the 'real', the presentation of a 'slice of life', an examination – in cross section – of fin-de-siècle society's hierarchies and conventions. Within these works – which were frequently performed and popular, aroused controversy, and were censored to varying degrees – an attempt to create the illusion of 'everyday conversation' was engaged. There are numerous staging options their translator might textually encode for a target culture theatre director, via her or his strategies concerning linguistic conventions associated with the 'real', 'contemporary', 'historical' and 'cultural other'. A theatre translator must try to imagine various potential approaches to the staging, design and casting of a play in her own context as well as taking into account previous productions. She must also try to hear potential multiple nuanced ways in which an actor might deliver a line, discover and develop subtext. She must try to catch and imprison the multiple theatrical possibilities she perceives in the play text, in the target language, whilst imaginatively negotiating her own dramatic and theatrical landscape. This is often why she sits in isolation, anxiously mouthing something to herself, before allowing her text to be read aloud by others. Theatre translation is as much about an impulse towards preservation, a sort of linguistic embalming, as it is about the potential for new embodiments.

Defining research methodologies

The research activities I have devised in order to engage with Zapolska's dramatic writing form six integrated methodological strands: firstly, via consideration of available critical material, the identification of a requisite historiographical research field to facilitate readings of fin-de-siècle, predominantly naturalist, theatrical practices in 'Poland'; secondly, archival research into the performance history and reception of Zapolska's plays. This presents complications, since no single archive exists holding the playwright's work. In addition, processes of partition, war and censorship have shaped its reception – as text and performance – in specific ways. Based on these investigations, I can begin an attempt to visualize previous staging solutions and speculate about contexts and theatrical conventions. The nature of the archival material available shapes these speculations; the apparent absence of particular documents and difficulties in accessing material is as significant as what is physically present. This is often fragile and poorly catalogued, but can be handled. Archival structures and processes are ideologically reflective and expressive. Consequently, this research is framed reflexively and this reflexivity informs decisions I subsequently make as a translator and director.

The third strand comprises pre-rehearsal, first-draft translation into English and the fourth involves their animation as research performances. As such, translation in its broader sense comes to constitute both a (fundamentally collective) research method and an outcome.

Research performances at Reading represent examples of what is referred to in the field of Theatre Studies as 'research through practice'. This involves the critical exploration of particular research questions or problems through workshops and/or the staging of a production/performance, which may have evolved from a written play text or a process of devising. Accordingly, a formalized, annual nine week slot is available each autumn term for extra-curricular, staff-led research projects of a practical nature, in which students can also become involved. This opportunity provides an arguably indispensable experimental forum for the theatre translator: with my cast I work on staging productions of my new translation, the rehearsal process becomes a way of developing the translated text, and this changes responsively week by week. A developmental, rehearsal-based working method of this kind is not dissimilar from that occasionally employed by the playwright herself, who was also, interestingly, a translator of plays, predominantly from the French.

The fifth research strand involves sometimes extensive re-fashioning of the text following the witnessing of public performances of its first 'incarnation', with particular focus on rhythm, tone, visual and verbal humour (absolutely crucial to and very complex in Zapolska's work) and specificity of 'voice' in the definition of character. I tend to engage in this re-fashioning following a few months' break, by returning to the Polish original, the contextual research I have amassed and, importantly, memories of actors' movements, gestures and voices during rehearsal and performance, as well as audience response. Traces of these performances (or sometimes readings or workshops) are inscribed to a large degree in the final published versions, which themselves are by no means, of course, 'fixed' or 'stable'.

Characterizing the sixth strand of my research is interrogation of how interactions between so-called source and target cultures and languages problematically manifests themselves as theatre, since my research performances are staged in spaces occupied by the academic community and in spaces occupied by what might be classed as the 'Polish community', both in the UK and abroad. In addition, as the daughter of Polish émigrés, the translation and direction of dramatic literature written in Polish involves extensive consideration, forming a fifth research strand, of my own work within narratives of emigration, deportation and exile from Poland and post-World War II processes of Polish/British identity formation.

Zapolska's corpus and its cultural transposition

Some of the challenges of working on the translation of Zapolska's plays can be expressed by referencing two plays she wrote in the 1890s – *Małka Szwarcenkopf* (1897) and *The Other* (1898). These are concerned with processes of translation and cultural difference. *Małka* is written predominantly in Polish and set in Warsaw. The play also contains sections written in a combination of Polish and Yiddish (and was also translated into Yiddish, which Zapolska reportedly could not speak) – most significantly a betrothal ceremony of dubious 'ethnographic authenticity'. The play is about arranged marriage and divorce in a Jewish community. The pressures the young protagonist Małka and her future husband Jojne are exposed to are explored by Zapolska. The play ends with Małka's suicide.

The Other is written predominantly in Polish, which is combined with Russian. In this play Zapolska takes up the topic of conspiratorial activity for the cause of Polish independence and the counter-active machinations of the military secret police in the Russian partition. The Russian Zapolska has used has been described as 'polonized'. The central character, a military policeman whose job it is to root out anti-imperial conspirators, is bilingual and bicultural. The fact that his mother is identified as Polish problematizes his professional activities, rendering him both particularly dangerous and particularly vulnerable. His ability to move seamlessly between the two languages and to 'own' them in a very particular way is used by Zapolska to anchor her exploration of themes of internal and political conflict, otherness, duplicity and performativity within power relations in both private and public environments. The politicized practice of translation in contexts where many characters understand both languages in question to varying degrees (partially accounted for by considerable phonetic correspondences) is a central feature and problem of the play. Both these dramas were hugely popular in their time – most specifically in Galicia, where government and censorship were by far the least aggressive. Any attempt to now translate and stage them raises significant semantic, and therefore political and ethical, issues. Bringing the translation process of these plays into the rehearsal room allows me to engage dynamically, spatially, and physically with questions about ideology and the ethics of theatrical representation. The relationship between actor and character that is implied within these naturalist play texts renders this process particularly important. Such a working method also facilitates the identification of potential strategies for animating these plays in a new context in such a way that the full import and context for, the relationships in question is at once expressed with immediacy *and* historicized. The main issue being: where and when, conceptually, spatially, physically, linguistically, within the here and now, can one 'locate', 'situate', 'map' these plays, particularly given the fact that many are being translated into, and performed in English for the first time, finding cultural and literary reference points can be difficult.

As well as writing plays which tackled questions of national and cultural identity, Zapolska wrote many works about the Polish bourgeoisie. Irena Krzywicka has described her as a writer who reveals the functioning of class systems by repeatedly 'opening up' within a theatrical space the domestic interiors of Krakow, Lwów and Warsaw, from the cellars below to the attics at the top, in an image akin to a doll's house. Krzywicka argues that Zapolska's overwhelming focus on relationships within the urban domestic sphere, predominantly a matriarchal space, held as implicit for its contemporary audience the acute political and social disempowerment of Polish men in the public domain during the years of occupation and repression. Krzywicka violently gives image to the Polish fin-de-siècle as a time of sadness, dirt and depression, encapsulated in the fashionable form of the contemporary woman, 'monstrously distorted by means of whalebone – her body, her neck, her soul. A suffocated stomach, a stiffened posture, innards pushed up into the ribcage, enormous breasts clamped into one roll, hips like millwheels' (Krzywicka, 1961). For Krzywicka this almost industrial structure epitomizes the 'unnatural'. She concludes that nothing good can be said of an epoch that thus shaped its women and posits Zapolska as the herald of social change – a herald, we may add, of revolution in the domestic sphere and most progressive in her representation of gender politics, notably in plays such as *The Man* (1902).

What strikes me about the *mise-en-scène* of this category of her plays is the proliferation of doors and windows. Energy is created by frequent entrances and exits, evocative of bedroom farce, slapstick and the comedy of manners. This is usually drawn into tension with an absolute intensity of focus on one interior, characteristic of contemporary Naturalism, as providing a structure for relationships and the territory where Polish 'domestic' politics are played out – predominantly in Polish – a language that had been under threat during the darkest days of the Russian and Prussian partitions. Further investigation may well be required into how precisely Zapolska's dramatization and theatricalization of this sphere might be read as constitutive of a sophisticated dialectic concerning the relationship between form and national politics. Norman Davies (1986), narrating Polish history of this period, comments that the scope for minority politics was extremely limited. Political deprivation was a major fact of life. Zapolska appears to question what precisely might be defined as political in nature. 'The personal' is usually categorized as the domestic and, it follows, feminine, private, unsuited to and distinctly separate from the public sphere. Accordingly, via the dramatic medium, Zapolska interrogates whether 'the personal' and 'the political' are necessarily incongruous and mutually exclusive concepts.

A brief research through practice case study

Employing an integrated set of methodologies that involve research through practice – in the forms of translation, a developmental rehearsal process and performance – allows for full exploration and eventual textual expression of these thematic strands and their correspondent aesthetic forms. It also encourages speculation – within the context of historical research – about fin-de-siècle perceptions of naturalistic theatrical space and its complex relationship with an elusive, censored socio-political 'reality'.

During my 2003/4 research performance of Zapolska's 1906 play, *The Morality of Mrs. Dulska*, for example, this thematic preoccupation with concepts of the personal and political, specifically within a domestic space, was extended extra-textually via the use of projection. Images of maps and of a variety of public and domestic buildings were juxtaposed rhythmically with similarly monochrome, and sometimes fragmented, abstract, images of period furniture and costume, echoing the appearance of the actual set and actors. These signs were chosen for their perceived similarity to each other, so that the pattern of a lace collar, for example, recalled the 'pattern' of a map, a section of window frame, the frontage of a tenement building recalled the structured shape of a woman's corset. In addition, whereas original productions would have employed a constructed and enclosed box set from which characters exited to that fictional place 'the rest of the house', my actors/characters were constantly in view of the audience, seated on chairs outside five large free-standing, decorative door frames, with their backs to the auditorium, when they were not 'performing' within the domestic environment. When actors entered the performance space, they were momentarily 'framed' by decorative door-frames. The resultant tableaux were suggestive of early portrait photography and pre-figured the behavioural stylization and conventionality inherent in performances. Framing this arrangement, though for the most part located in darkness, were the permanent architectural features of the black-box theatre itself.

Of the texts Zapolska developed for performance *The Morality of Mrs. Dulska* (1906) is probably her best known. The play reached practically canonical status merely a few years after it was written. To this day its largely unresolved status as both a popular work intended for performance and a classic literary text worthy of academic study endures. This is maintained by critical discourses which tend to re-enforce polarization within a high art/low art value system. However, this status has been considerably problematized by vigorously politicized attitudes towards the playwright's sex and her biography. This example of 'woman's work' occupying an atypical space within critical and cultural discourse has sporadically provoked intense scrutiny of the relationship between Zapolska's literary and personal reputation. Arguably, this in turn has ensured the play's enduring presence on Polish stages. The wily, tyrannical, petty-bourgeois landlady, Mrs. Aniela Dulska, and her not-completely-silent husband, long-suffering Felicjan, have been persistently referenced and recreated in the sphere of Polish cultural production and memory. The new play was rapidly snapped up by permanent theatres and touring companies. It became an immediate hit with audiences across partitioned Poland. No surprise, then, that the play provided the source and its title, the root, for a new term in the Polish language: 'dulszczyzna', roughly translatable as 'dulskaness'. This has been employed as a 'catch-all' for the litany of reprehensible qualities exhibited by bourgeois philistine, Aniela Dulska, who effectively pimps her serving maid to her son in order to 'keep him at home'; double standards, endemic conservatism, excessive self-delusion, poor social conscience, weakness of character, hypocrisy, xenophobia, penny-pinching, vanity, pomposity, crassness, lack of compassion, sadistic self-aggrandizement and bad taste. These qualities have been variously described by critics as indigenous to Poland, foreign or universal. Correspondingly, the complex ideological tensions provoked by this linguistic phenomenon have served to further the so-called cultural immortalization of the text. Like characters from a TV soap opera who receive letters from the viewers, the almost-real Dulskis have become a supra-theatrical embodiment of capitalist miserliness and secretly relished moral dysfunction.

My research performance of this play was staged at the Centre for Polish Culture (POSK) in Hammersmith, London. The three public performances followed on from a previous run which had taken place at Reading University in autumn 2003. POSK is a fully-equipped theatre venue at the heart of London's so-called Polish community and was established by post-war Polish émigrés. It regularly hosts professional companies from Poland as well as more local events. Pupils at many of London's Polish Saturday schools had at the time been reading the play, which featured on their exam syllabus. Consequently, the event was billed as part of a broader educational and fund-raising programme; an opportunity for students to see a play they were studying in Polish, performed in English. Thus, the project also came to represent a very particular instance of outreach between two groups of students and two communities. On one very lively and, for me, unforgettable afternoon the production played, in a theatre packed to the brim, to over three hundred vocal and energetic Saturday school students, aged roughly between six and eighteen, many of whom cheered loudly and applauded when the Dulski's wronged (and predictably pregnant) servant girl, Hanka, demanded her one thousand kronen from Mrs. Dulska. Some of them are the children of recent immigrants, whose mother tongue is Polish. Some are the descendants of the post-war diaspora and those who left Poland, in much fewer numbers, during the latter half of the twentieth century. For these latter two groups, and I include myself in the first, Polish language acquisition may be proving increasingly challenging.

Such Saturday schools were established after World War II, predominantly by political refugees, all over the UK and have been maintained with great passion and dedication. Following Poland's accession to the EU, they are facing the renewed challenge of self-definition. They strive once again to engage in dialogue concerning the philosophical basis for their pedagogical approach and, indeed, their existence. This dialogue must surely respond to the shifting demographic trends now having an enormous impact on the make-up of their staff and student body, for whom concepts of 'difference' and 'otherness' are undergoing new, not unwelcome, forms of de-stabilization. In order to survive, these schools will have to accommodate and develop vastly varying language abilities and attitudes towards cultural heritage. They will be compelled to hear, and respond to, the growing number of voices now contributing towards currently evolving debates concerning identity politics and multiculturalism in the UK.

In developing a register for this particular translation, a process ultimately enabled by the research performance, several different factors were taken into account. I aimed for formality of address, in order to effect a 'historicization' of the action, judging that too contemporary a tone would fail to evoke the late nineteenth-/early twentieth-century milieu and thus defuse the scandal that takes place in the Dulski household and is central to the play's action. Zapolska wrote the play a century ago and so I aimed to create a 'linguistic construction' of that period in English. The formality of address can be employed variously, especially by the actor playing the main character, Mrs. Dulska, to demonstrate, among other qualities and strategies, class aspiration and/or social sophistication via enabling the expression of varying degrees and methods of politeness and affectation and a constant negotiation between the performed public and private 'selves'. All these concerns are central to the Polish text, in which the evocation of the Dulski family's double standards is crucial. In addition to aiming for a formal quality that implies historical distance, I also attempted to locate a register which fuses, from a UK perspective, implied 'otherness' of location with patterns and rhythms of speech that conform, where appropriate, to perhaps more contemporary British stereotypes of the middle class, in the hope of Zapolska's satiric purpose being more readily realized in performance.

The following extract shows Zapolska telling her tenant that she can no longer offer her accommodation.

Dulska: A dreadful affair . . . your husband . . . all the business with that girl . . . that is one thing . . .
Tenant: But she was my servant. It was revolting! I couldn't bear it! As soon as I knew for sure, I . . .
Dulska: You swallowed some match heads. Such second-rate poison! People were laughing. Nor was that the end of it. The whole affair was just like a comedy! If you'd at least died . . . well then . . . !
Tenant: It is a great pity.

Dulska:	That is not my implication – only death is always something... there is always something to it... but that... well... all I can say, dear lady, is that everyone laughed heartily. On one particular occasion, I was travelling home by tramcar. We passed my tenement building, because the stop is a little further on, and two gentlemen – I didn't even know them – indicated the building saying, 'Look! There is the house where that jealous wife poisoned herself' and they began to laugh. I thought I would freeze to the spot on that tram.
Tenant:	*(Humbly)* I am truly sorry for all the unpleasantness I've put you through.
Dulska:	Oh, my good woman, publicity is publicity...
Tenant:	The whole incident made me very ill. Besides which, I didn't know what I was doing. I was driven, like a madwoman then... *(She weeps quietly)*
Dulska:	Of course you were, my dear. Each and every suicide must be mad, must have lost her sense of morality, her faith in the omnipresence of God. That's true cowardice if ever I saw it. Quite so, in a nutshell – cowardice! Followed by the annihilation of the soul. It is a great comfort to know that suicides are buried in seclusion. They should not be permitted to push and shove their way into the company of decent people. Kill yourself!... And for whom? For a man! No man, my good woman, is worth risking eternal damnation for.
Tenant:	Pardon me, but this was not about a man, it was about my husband.
Dulska:	Huh!
Tenant:	I was not able to tolerate *that* under my own roof.
Dulska:	Better under your own roof than someone else's – less publicity. No one need know.

The setting for *The Morality of Mrs. Dulska* is unquestionably Galicia, detectable from the currency referred to (gulden, krone) as well as terminology employed by members of the household, very difficult to render in translation. The setting for two of Zapolska's spin-off short stories, *The Death of Felicjan Dulski* and *Mrs. Dulska in Court* (1907), is unambiguously Lwów, or Austrian Lemberg. However, writing in 1995, Kowalski suggests that, for a post-1945 spectator of the play in possession of a modicum of precise geographical knowledge, a rather bizarre evocation of place may have emerged. He asserts that this was the result of politically motivated interference by the censor. Though landmarks and street names referred to indicated the urban context to be Kraków, Galicia's second most significant centre, mentioned institutions suggested that it was Lwów, the capital of Galicia in 1906. The destination of Felicjan Dulski's 'stroll' was originally intended to be Castle Hill in Lwów. However, there is also a large castle on a hill in Kraków. It was also in Lwów, in 1869, that, with the help of public funds, a mound was raised to commemorate the 300th anniversary of the Union of Lublin, a joining of political forces with Lithuania. Conveniently, there is also a landmark mound in Kraków, the Kosciuszko Mound erected in the early 1820s. After 1945, however, Lwów, as Lvov, was part of the Ukranian Soviet Socialist Republic and, Kowalski implies, attempts were made to suppress its Polish identity. This involved the censorship of key texts, such as *The Morality of Mrs. Dulska*, in which the play's action was 'mapped' onto Kraków, still part of Poland. The city featured sufficient similarities to carry the change. Resultant 'slippage', however, could not entirely be avoided, the outcome an uneasy conflation of two different places. Best not to dwell on the possible existence of a nascent bourgeoisie (the Dulski family) and a couple of ineffectual proletarians (the pregnant serving girl who asks for a large pay off and her wily, persuasive godmother) in post-war Lvov, implies Kowalski (1995). To add to the apparent complication, Czachowska, writing roughly thirty years *earlier*, provides information relating to the play's performance history not mentioned by Kowalski. Though Zapolska set the action in Lwów, Ludwig Solski, director of the play's premiere in Kraków, 'transferred' it, in the interests of relevance and immediacy, to Kraków, changing requisite references accordingly. When the text was published, in 1907, the location, seemingly in accordance with Zapolska's intentions, featured as Lwów. The 1924 edition, according to Czachowska (1966), followed Solski's 1906 prompt copy and returned the location to Kraków, where it remained until the subsequent 1958 edition. An earlier instance of censorship relating to shifting territorial identities appears to be implied, in response to the Polish-Ukranian conflict over Eastern Galicia following the collapse of the Habsburg Empire in 1918. Both Ukranian and Polish victims of this conflict are buried in Lychakivsky cemetery, in what is now L'viv, as is Zapolska. The Polish-Ukrainian-Soviet-Austro-Hungarian struggle for Eastern Galicia, historically a passionately contested area, was repeatedly re-inscribed in the 'geographical' vacillation of the Dulski family home.

Preparing to stage the play in Reading in English, I decided to locate the action in an imagined nineteenth-century Kraków. This was significant due to certain decisions regarding *mise-en-scène*. The use of entr'acte patterns of projections covering the entire high wall at the back of the large stage space allowed me to unambiguously signal a semi-fictional place. The projections were monochrome and spilled onto and over what was effectively a black-box theatre space, filled with minimalist though period-specific elements of setting, including furniture, an indispensable samovar and five free-standing door frames. The set formed the 'skeleton' of an early-twentieth-century middle-class drawing room whose various elements were regularly, almost symmetrically, arranged to form an open spatial relationship with the audience. This encouraged actors to direct performance outwards, displaying the accoutrements of 'character' (costume, facial expression and gesture) without understatement and occasionally with stylization. Among the projected images employed, as I have already mentioned, were pre- and early twentieth-century photographs, woodcuts and prints of Kraków maps and landmarks. My decision regarding 'place' arose from the judgment that an audience not necessarily conversant with Polish history and culture would more readily recognize Kraków as Polish. To this extent, I perhaps inadvertently colluded with the intentions of the post-1945 censor mentioned by Kowalski (1995) and created even greater 'slippage' than strictly helpful when staging a translated text for research purposes.

Once the production transferred to POSK – the Polish theatre in London – where it also played on two occasions to a more uniformly British-Polish audience, predominantly of the older post-war generation – all of whom speak fluent English – my decision suddenly acquired new meaning – political meaning – and prompted a few raised eyebrows. The sense of nostalgia for Polish Lwów – now Ukranian L'viv – is still strong. The play had received its premiere in Kraków, in 1906, so the geographically specific images were occasionally rationalized from this perspective. However, my justifications concerning familiarity and strangeness inevitably floundered at this juncture. Nor could I bring myself to tackle head on the still somewhat sticky point, as far as much-loved, well-known Polish 'classics' are concerned, of so-called directorial artistic license and its relationship to theatre as a live event. My decision to wryly accompany the entr'acte projections with extracts from a mid-nineteenth-century Polish opera entitled *The Haunted Manor (Straszny Dwór)* by Stanisław Moniuszko, building on one of the character's love of all things operatic, in addition evoked very different readings in both performance contexts, depending on the audience's understanding of the libretto and consequent attempts to engage inter-textual connections. For the Polish speaker, the opera could have been read as significantly more than musical accompaniment governing the rhythm of a series of projections, because they may have been able to locate it culturally, reading in addition between the two texts they were hearing (the recorded, sung and the live, spoken) on multiple levels. Inevitably, in each performance context, some aspect of my intentions was rendered to some extent invisible – even, potentially, read as unsystematic or very specifically politicized – given the viewer's interaction with the signs in question and directorial decisions concerning target audience. In my published translation, as in all late-twentieth/early-twenty-first-century Polish editions of the play, the setting is Polish Lwów – or, perhaps depending on one's perspective, Austrian Lemberg of 1906. In 2006, this city is in Ukraine and is called L'viv. In my translation, character names and place names have not on the whole been 'domesticated'. However, the linguistic patterning of this translation also implies that the action is taking place in England, somewhere between the present and an imaginative construction of the early twentieth century. Watching my own production, my sense of being suspended in a sort of spatial, temporal and linguistic 'no-woman's-land' was acute. This was most certainly a *physical* sensation. I was in a very lonely – and a liminal – 'place'.

Conclusion

Lonely or not, there is a huge amount of work to be done in terms of framing and theorizing my own research, refining my expression of ideas and the coherent development and conceptual integration of translation and performance practice within my chosen mode of academic investigation. It would be my wish to see a performance of one of my translations of Zapolska's dramas on a professional English-speaking stage, playing to a range of audiences, including perhaps émigré Poles. Although the dissemination of my work in print clearly provides me with a huge amount of satisfaction, it seems more fitting that practice as research should generate further practice, in a range of contexts, both in terms of performance and re-translation. This is a responsibility I take very seriously, given my cultural background. As a translator, I am creating some of the conditions that might enable this to happen. Whether it will or not depends on the reception of Zapolska's work and the effectiveness of my interpretative, impassioned mediation.

References

Czachowska, J. (1966) *Gabriela Zapolska: Monografia bio-bibliograficzna* (Kraków: Wydawnictwo Literackie)

Davies, N. (1986) *Heart of Europe: A Short History of Poland* (Oxford: Oxford University Press)

Kowalski, M. (1995) 'Przeprowadzka państwa Dulskich', *Glos* (35:1), p.20

Krzywicka, H. (1961–2) 'Polska Komedia Ludowa', *Listy Teatru Polskiego* (56), pp.11–15

Murjas, T. (ed. and trans.) (2007) *The Morality of Mrs. Dulska by Gabriela Zapolska* (Bristol/Chicago: Intellect)

Murjas, T. (ed. and trans.) (2009) *Zapolska's Women: Three Plays* (Bristol/Chicago: Intellect)

Index

Page numbers in **bold** indicate figures and tables.

Aaltonen, Sirkku, 244
Abraham, Natalie, 207
Acco Festival for alternative Israeli theatre, 218, 224
A Jia, 110, 111
Amit-Kochavi, Hannah, 219
Anouilh, Jean, *Antigone*, 87, 93–100
Arabic-to-Hebrew translation, *The Impotents* (Masarwy), 6–7, 215–27
  Acco Festival, 218, 221
  Arabic performance with Hebrew surtitles, 218, 221
  audience imposition of conflictual significance, 7, 223–6: Arabic speakers, 224; Hebrew speakers, 224–5; interpretation of surtitles, 225–6
  available translators, 219–20
  Cameri Theatre reading, 218
  conflict-related translations, 7, 215–16, 226–7
  directed by author, 218
  effects of ongoing conflict, 220
  ex-military translators, 219–20
  imposition of political significance, 216, 218–20
  Israeli-Palestinian conflict, 218
  linguistic characterisation of dialogue, 221–2
  literal translation, 222–3
  multi-phased process, 217–18
  play based on short story, 217
  play choice, political nature of, 218–19
  Tel Aviv University theatre studies, 217–18
  theatre as a political tool, 223
  translational decisions, 220–3: achieving performability, 220–1; resulting archaism, 223

Artaud, Antonin, 236
Aufricht, Ernst, 127–8
Avon Foundation Writer centre, Shropshire, 201

Babel, Isaak, *Marya*, 173, 184
Bailey, Mark, 126, 133
Baker, Mona, 215–16, 226–7
Baker, Tim, production of *Threepenny Opera*, 126, 130–2, 134, 137
Balas, Shimon, 219
Bal, Mieke, 18
Bamborough, Karen, 179
Barker, Howard, 244
Bassnett, Susan, 157, 237
  on performability, 14
  and Trivedi, Harish, *Post-Colonial Translation*, 232, 234
Beckett, Samuel, *Waiting for Godot*, 32, 140
Berliner Ensemble, 127, 129
Berman, Antoine, 60
Bernhard, Thomas, 184
Blackdrop performance poetry events, 64
Blair, Rhonda, 107, 122
Blakemore, Diana, 91
Blitzstein, Marc, 129
Blom, Eric, 128–9
Boeglin, Bruno, 64
Bogdanov, Michael, production of Brecht, 130
Bolt, Ranjit, 196
Bond, Edward, 174
Boswell, Lawrence, 207
Bourdieu, Pierre, 156
Bourne, Matthew, 189
Brault, Christophe, 245
  and Dias, Jean-Paul, 230
Bray, Barbara, translation of Anouilh's *Antigone*, 93–8

Brecht, Bertolt, 127
  at the National Theatre, 192
  audience involvement, 131
  Joe Fleischhacker, 128
  Lehrstücke, 131
  Mahogonny Songspiel, 128, 132
  'On Chinese Acting', 110
  performing his work, 135
  problems with translation rights, 5, 126, 180
  see also The Threepenny Opera (Brecht)
British Grotowski Project, University of Kent, 250
British Sign Language (BSL), stage interpretation, 4, 72–85
  actor–audience relationship, 78
  basic requirements of translation, 79
  complex assignment model, 79–80, 79, 85
  depth and detail of translation, 83
  distraction to hearing audience, 84
  dramatic dialogue, 75, 77, 85
  Ibsen's *Hedda Gabler*, 77–8, 84
  interaction of translation/ interpretation/performance, 76, 76
  interpreter: and theatre practitioner, 84, 85: theatrical competence, 83; and theatrical élan, 80; training, 74–5, 85
  markers of plot/character/ relationships, 81–2
  multi-modality of audiovisual communication, 76–7
  preparation for assignment, 74, 80
  scene as unit of translation, 80
  Shakespeare's *Richard III*, 78
  Sing Yer Heart Out for the Lads, 80–3
  translation of *theatrical* text, 75–6
  unnecessary dialogue, 77, 78
  visual elements, 77–8, 82, 85
British Theatre, translation work, 197–9
  see also English theatre and non-English language texts
British theatrical system, cultural imperialism, 231–6, 244
  audience preferences, 233, 234, 244
  commercial imperative, 232, 244
  dangers of clarity and realism, 244
  domestication of dialogue, 235
  foreign drama marginalised, 231, 233–5, 244
  linguistic/cultural specificity, 244
  naturalistic style, 233, 234–5, 236
  performability and marketability, 231–2, 233
  self-censorship, 232–3
  stage translation, 231, 233–6
Brown, Kenneth, 158
Brownlie, S., 227
Büchner, Georg, 238
Bulgakov, Mikhail
  Crimson Island, 206
  Flight, 185

Calderón
  El pintor de su deshonra, 26–8
  translation of, 27–8
Canada
  linguistic territories and minorities, 155
  Quebec and Francophone identity, 160
  see also Cow-boy poétré; Western Canada, bilingual drama
Castledine, Annie, 129
Channels project (National Theatre Studio), 197, 201–2, 211
Chefitz, Eric, 234
Chekhov, Anton
  at the National Theatre, 192
  translations of, 14–15, 174, 176
  writing practice, 175
Chen Shizheng, 118, 119, 120, 121
Chéreau, Patrice, 51
Chernoff, John Miller, 58
Chikmatsu, 198
Chinese *xiqu*, 108–22
  actor–character–audience relationships, 110–12
  actor training, 109
  adaptation, 108, 117
  embodied experience, 109
  fundamental aesthetic, 114
  Jingju, 109, 120, 121, 122
  kunqu, 109, 118, 119, 120

Mei Lanfang, 110
zaju, 109, 117
see also embodied approach to translation in Chinese *xiqu*; embodied approaches to *xiqu* adaptation
Cobbe, Rosie, 195
collaborative theatre translation projects, 200–10
　Channels project (National Theatre Studio), 197, 201–2, 211
　The Fence, 6, 200–1, 204
　The Gate Theatre, 205–7: British premieres, 206; collaborative process, 205–6; commodification of international work, 207–8; home-grown international theatre, 205; international policy, 197, 206; Spanish Golden Age, 207; translation practices, 206–7; transnational identity, 207
　Janus, international cultural project, 6, 200, 201
　Traverse Theatre, Edinburgh: audience response, 205: issues of copyright, 210; new and international work, 202, 203; Playwrights in Partnership, 204–5, 209–10; role of literal translator, 208–10; translation and collaboration, 203–4
　use of literal translations, 202
　West Yorkshire Playhouse, 202
　Writernet, 6, 200
　writer-to-writer involvement, 202
Connery, Sean, 183
Cottesloe studio theatre, National Theatre, 188, 189
　Education Department, 191
Cournoyer, Daniel, 158
Coward, Noel, 191
Cow-boy poétré, creation of a bilingual play, 158–67
　audience specific adaptations, 164–7, **166**
　English-French translation, 158–9
　linguistic duality, 161, 167
　original English text, 158
　presentation at Ottawa festival, 160
　regional differences in French dialogue, 158–9
　rodeo background, 158, 161, 162
　tour of Western Canada, 160–1
　translation for francophone audience, 161–7: incorporation within actors' speeches, 163–5; use of surtitles, 161, 162
Cracknell, Carrie, 207
　directing *Mobile Thriller*, 148, 149, 150, 152
Craig, Ryan, 252
Croft, Giles, 206
Cronin, Michael, 232
Curry, Tim, as Macheath, 130
Czachowska, J., *Gabriela Zapolska*, 263

Daldry, Stephen, 206–7
Dalpé, Jean Marc, French-Canadian drama, 157
Darin, Bobby, 135
Davies, Howard, 183
Davies, Norman, 253, 258
the Deaf community, 72–3
　see also British Sign Language (BSL), stage interpretation
De Filippo, Eduardo, 140, 192, 198
Delgado, Maria, and Fancy, David, 49
De Marinis, M., 41
Demtraki, Stavros, 243
Dias, Jean-Paul, 230
Din, Ayub Khan, *Rafta Rafta*, 196
Dingwaney, Anuradha, 234
directing and translating, Maeterlinck's *Aveugles*, 31–47
　the director (Pavis), 46
　displacement of verbal text, 39
　focus on sonority and rhythm, 4, 33–5
　knowingness of the actors, 36, 37
　marking the boundaries of silence, 34–5
　metaphysical resonances, 34, 35–7
　non-mimetic live sound, 36, 37, 38–9
　non-verbal elements of performance, 34, 35–7, 39
　physicality of performance, 39–40

directing and translating – *continued*
  texts, written, and performance, 40–3
  theatrical discourse/dramatic action, 4, 39
  translator as 'metteur en scène', 4, 31–3
  use of silence, 36–8
  verbal/non-verbal signifying systems, 33
  writing a performance text, 4, 36–47, 40–7: development through rehearsal, 44, 46; gaps filled by performance, 47; 'gaps' in the written text, 40; residue text or director's metatext, 40–3; roles of translator and director, 44–6
Dodgson, Elyse, 197
Dresher, Paul, 118, 119, 120
Dürrenmatt, Friedrich, 29

Early, Michael, 231–2
Eddershaw, Margaret, 130, 132
Edinburgh Festival Fringe (2004), production of *Mobile Thriller*, 149–52
Edis, Steven, 126
embodied approaches to *xiqu*, adaptation of *Injustice Done to Dou E*, 117–21
  American Repertory Theatre (ART) production, 118–20
  critical reception, 119, 120, 121
  jingju basic training, 120, 121, 122
  kunqu arias and performance skills, 118, 119, 120
  original script, 117–19
  rehearsal tool for student performers, 120–1
  as *Snow in Sweet Summer*, 120–1, 122
  use of professional *xiqu* performer, 119–20
  Western experimental music-theatre, 119
  xiqu performers and techniques, 118
  xiqu principles, 120–1, 122

embodied approach to translation in Chinese *xiqu*, 112–17
  jingju performance techniques, 112–13, 122
  jingju vocal technique, 113–14, 116
  Judge Bao and the Case of Qin Xianglian, 113, 114–15
  'playable' translation of *jingju* performance convention, 113
  preparatory physical training, 114
  rehearsal script, 114, **115**, 120
  vowel sounds, Mandarin and English, 114, 116
English language audiences, attitude to foreign plays (Hampton), 184–6
English theatre
  London's West End, 193–4
  Royal Shakespeare Co., 193
  see also National Theatre (UK)
English theatre and non-English language texts, 6, 187, 194, 199
  play selection, 186, 194–9: drama for export, 195; playwriting in Britain and Europe, 195; power of the director, 195, 198; production of foreign work, 196
  see also National Theatre (UK)
Erdman, *The Mandate* and *The Suicide*, 198
Erskine, James, 148
Euripides, *The Trojan Women*, performed in Israel, 223–4
Eyre, Richard, 187, 198

Fancy, David, 49
Farquhar, *The Recruiting Officer*, 31
Farr, David, 207
Fiennes, Ralph, 181
Fish, Stanley, 12
Fitzgerald, Susan, 15
Fort, Paul, 32
Fosse, Jon, 192
Franzon, Johan, 60
Frayn, Michael, 14, 15
  translation of Chekhov, 176
French contemporary theatre
  actors must accommodate text, 238
  and the British establishment, 244

contemporary dramatists, 230–1
experimental nature of, 231
experimentation in language and
  form, 236–7
'irreducibility of the Poet', 238
language to 'produce' not represent,
  236
poetic opacity of language, 235–6
primacy of language, 230–1
see also Renaude, Noëlle *Par Les
  Routes*
French contemporary writing, and
  translation, 235
Freytag, Holk, 224
Friel, Brian, 15
Froud, Nina, 174
Fuegi, John, 127

Gabrielli, Renato
  Qualcosa Trilla, 5, 139, 140
  Vocation and *The Number Ninety's
    Child*, 148
  work on *Mobile Thriller*, 149, 150,
    151
  see also Mobile Thriller
Gao Xingjian, 110
Gardzienice Company, 250, 251
Gareau, Laurier, 158
The Gate Theatre, 205–7
  British premieres, 206
  home-grown international theatre,
    205
  international policy, 206
  international work, 197, 203, 232
Gay, John, *The Beggar's Opera*, 128
Gill, Peter, 174
Glenny, Michael, 173, 174, 184
Gooch, Steve, 245
Gordon, Mick, 203, 207, 232
Granville-Barker, Harley, 188, 191
Grégoire, Hélène, 174, 175
Greig, David, 205
Grice, Paul
  'Logic and conversation', 88–90
  theory of conversational
    implicature, 87–90, 92–3
Griffiths, Trevor, 179
Grotowski, Jerzy, 250, 251
Grutman, Rainier, 156

Guang Hanqing
  Injustice Done to Dou E, 117–19
  Yuan Dynasty *zaju* play, 117, *see
    also* embodied approaches to *xiqu*
Gupta, Tanika, 196

Hale, Terry, 244
Hall, Peter, 176, 179, 193
Hamberg, Lars, 237
Hampton, Christopher, translating
  and adapting, 6, 173–86
  actor input, 174, 175, 179
  adaptation, misleading term, 184
  affinity with the author, 176–7,
    178, 182
  archaic language and modern
    idiom, 178–9
  assessing a translation, 183–4
  Chekhov, interpretations of, 176,
    177
  English language audiences and
    foreign plays, 184–6
  Hedda Gabler, 174–5, 177
  Horváth, 178, 180
  Ibsen, interpretations of, 174–5,
    176, 177
  Laclos' *Les Liaisons Dangereuses*, 178
  Molière, 177
  patterns and signs in the original
    text, 175, 176
  relationship with translation, 176
  respect for the original text, 174,
    175, 176–7
  the Royal Court Theatre, 174
  the skill of the playwright, 180–1
  status of the translator, 184
  The Seagull, 174, 175, 178, 179, 185
  The Wild Duck for the National
    Theatre, 179, 180
  translating from a known language,
    177
  Uncle Vanya, 174, 175, 176, 178
  writing from literal translation, 6,
    174, 177
Yasmina Reza, 181–3: *Art*, 181,
  182, 183, 184, 204, 233–4;
  contemporary author, 182–3;
  English and American versions,
  181

Handelzalts, Michael (2005), 225
Hare, David, 189
Harrower, David, 205
Hart, Mickey, 58
Hauptmann, Elizabeth, translation of Gay's *Beggar's Opera*, 128
Henson, Basil, 179
Hickey, Leo, translation of Lorca, 24, 25
Higgins, David *see* 'Stickman', David
'Stickman' Higgins
Hinton, Stephen, 130
Hodgart, Hugh, 141
Holt, Marion Peter, 13
Horváth, Ödön von, 191
  performances of, 185
  plays, Hampton's translations, 178, 180
  Tales from the Vienna Woods, 180, 193
House, Juliane, 91
Howard, Philip, 149
Hughes, Ted, translation of Lorca, 24
Hytner, Nicholas, 193, 196

Ibsen, Henrik
  Hedda Gabler, 77–8
  National Theatre productions, 192
  see also Hampton, Christopher
implicature and inference in staging translation, 4–5, 87–102
  Anouilh's *Antigone*, 87, 93–100
  context, co-text, translated texts, 90–2
  Grice, theory of conversational implicature, 87–90, 92–3, 93–8
  inferences, source to target text, 93–8: character of Creon, 97–8; character of the Nurse, 97; flouting of the maxim of quality, 96–8; present tense/past action, 95–6; subtle hierarchy of meanings, 94–5
  staging process, recovery of lost implicatures, 98–102: actor's diction, 100; directorial adjustments, 99–100; gap between text and situational context, 100–2; Pavis' 'mise en scène', 100–1
Informal European Theatre Meeting (IETM), 201
Italian theatre, 139–40
  translator and the production process, 140

Jackson, Shannon, 11
Jarry, Alfred, *Ubu Roi*, 252
Jellicoe, Ann, 174
Jiangsu Province *Jingju* Company, 112–17
Johnston, David, 230, 231
Jourdheuil, Jean, 238

Kane, Sarah, 194
Kearney, Richard, 20
Kennelly, Brendan, translation of Lorca's *Blood Wedding*, 24, 25, 26
Kidd, Robert, 173
Klammmer, translations of Villon, 128
Koltès, Bernard-Marie, 49, 193
  African and Afro-Caribbean culture, 64
  Dans la solitude des champs de coton, 49
  reception on the English language stage, 49
  Une part de ma vie, 50
  see also Solitude (Koltès)
Kowalski, M. on *The Morality of Mrs. Dulska*, 262–3, 264
Krzywicka, Irena, on Zapolska, 257–8

Laermans, Rudi, 38
Lailey, James, 133
Landers, Clifford, translation theory, 14, 15, 237
Lasdun, Denys, 188
Lassalle, Jacques, 232, 243
Lefevere, 232
Le Moine, Philippe, 197, 201, 203, 204
Lenya, Lotte, 135, 136
Lepage, Robert, *Seven Streams of the River Ota*, 162
literal translation, 6, 174, 177, 202, 206, 208–10, 222–3

Li Zhenghua, 116
Lorca, F. García, 28
  Bodas de sangre (Blood Wedding):
    knife image, 24–6: translations
    compared, 23–6
    knife motif, 25
  The House of Bernarda Alba, 21
Lugné-Poe, 32
Lyons, J., 87
Lyttleton stage at the National
  Theatre, 188, 189, 196

McDonagh, Martin, *The Beauty Queen of Leenane*, 21
McDonald, Alex, 79, 81
McGuinness, Patrick, 39
McHugh, Greg, 148, 150–1
Maeterlinck, Maurice
  Les Aveugles, The Blind, 31–2
  physicality of performance, 39–40
  'second-degree dialogue', 35
  on silence, 37–8
  symbolist concepts, 32
  see also directing and translating Maeterlinck's *Aveugles*
Makeben, Theo, 128
Malmkjaer, Kirsten, 87, 91, 92
Mamet, David, 176
Mansfield, Susan, 148–9
Maragnani, Frédéric, 237
Marivaux, 192
Marowitz, Charles, 185
Marsh, Elizabeth, in *Threepenny Opera*, 135, **136**
Masarwy, Riad, *The Impotents: The Powerless Folk*, 215–27
Mayakovsky, *Bedbug*, 206
Mee, Charles
  Utopian Highway, inspired by Chinese original by Guan Hanqing, 118–21
  see also embodied approaches to *xiqu*
Mei Langfang, 113
Meschonnic, Henri, 51
Meyer, Michael, 180
Miall, David S., 20, 28
Miller, Arthur, 192
Miller, Jonathan, 185

Mitchell, Roger, 185
Mnouchkine, 196
Mobile Thriller (Gabrielli's *Qualcosa Trilla*), translator as cultural mediator and promoter, 140, 142–53
  Carrie Cracknell, director, 148, 149, 150, 152
  comparing texts, 143–8: characterisation of Massimo, 143–5; humour, 146–8; speakable dialogue, 145–6
  Edinburgh Festival Fringe, 5, 149–52: creative team, 149–52; critical reception, 151, 152; Herald Angel award, 151; innovative theatre venue, 149–50; inter-city tour, 5, 151, 152; sound track, 150; stand-up comedy and Scots idiom, 151; translation and co-adaption, 152
  Hush Productions, 148
  involvement of author, 149, 150, 151
  reading for *Scambiare*, 148
  Tron Theatre, Glasgow, 148
Molière, 28, 177, 178
  L'Avare (The Miser), translation with contemporary overtones, 22–3
Morahan, Christopher, 179
Müller, Heiner, 238
Murjas, Teresa, *Invisible Country: Four fin-de-siècle Polish plays* (2010), 250

National Theatre (UK), 187–90, 204, 232, 234, 252
  Education Department, 126, 131–7
  the Loft, 193
  the repertoire, 189, 190–1: adaptations, 191–2; foreign language translations, 192–3; Greek drama, 192; musicals, 191; revivals of British and American classics, 192; Shakespeare, 191
  Studio, 196, 198: Channels project, 201–2, 211
  Transformation Season, 193
  translation and adaptation, 187
  translation policy, 197

Neher, Caspar, 129
Neves, Josélia, 77
Nicol, Patricia, 151
Nida, E.A., 79
Nord, Christiane, 227
Novarina, Valère, 236, 238
Nunn, Trevor, 176, 187, 196

Olivier stage, National Theatre, 188–9
O'Loughlin, Kevin, 243
Oswald, Peter, 198

Page, Anthony, 174
Pavis, Patrice, 40, 41, 42, 100–1, 206, 230
 description of the director, 46
performability, 13–15, 18
performance-based translation *see* *Solitude* (Koltès)
Perteghella, Manuela, 230, 231
Perzyński, Włodzimierz, *Ashanti Girl*, 250
Pilot Theatre, 81
Plamondon, Crystal, 159
Poland
 Galicia: censorship and boundaries, 263: government and censorship, 257
 partition of (1795), 252, 253–4, 256–8
Polish community in Britain
 Centre for Polish Culture (POSK), London, 260, 264
 Saturday schools, 260: identity politics and multiculturalism in UK, 261
Polish theatre, 249–54
 in Britain, 250–2
 current deconstructive approach, 250
 effects of repression, 253
 fin-de-siècle playwrights, 250
 historically conditioned factors, 250–1
 impact of censorship, 250, 253, 263
 lack of translated primary texts, 251–2
 naturalism and realism, 250, 251
 naturalist playwrights, 253–4
 partition of Poland (1795), 252, 253–4
 role of the play text, 250
 see also translating Zapolska; Zapolska, Gabriela
practice and collaboration in theatre translation, 2–3, 7, 139
 see also theatre translation practice
Price, Gerald, 131

Qian Yi, 119, 121

Ramsay, Peggy, 126, 183, 198
Ramsden, Herbert, translation of Lorca, 24, 25
Ravenhill, Mark, 194
Renaude, Noëlle, 237, 245
 Des Tulipes, 237
 experiments in language and theatrical form, 236–7
 Fictions d'Hiver, 237
 Le Renard du Nord, 234
 Madam Ka, 237
 Ma Solange comment t'écrire mon désastre, Alex Roux, 236
 Par les routes, 230, 231, 237
 presents difficulties for directors and actors, 237–8
 reinventing theatre-making, 237
 translations of, 234
Renaude, Noëlle, *Par les routes*, politics of translating, 7, 231–45
 British theatrical system, 231, 232: primacy of financial concerns, 232
 By the Way, English translation of *Par les routes*, 238
 collaboration with author/ identification of linguistic features, 240–3: significance of sound and rhythm, 242–3
 collaboration with director and actors, 243
 Edinburgh Fringe, 231
 linguistic approach, 238
 linguistic translation, 231, 243
 maintaining the opacity of language, 235–6
 Renaude's linguistic coherence, 240

'stage' translation, 231–2
translation as a political act, 243–4
urbanised world of visual signs, 239
Reza, Yasmina, 178, 181–3
  Art, 181, 182, 183, 184, 204, 233–4
  Life x 3 , 181, 192
  see also Hampton, Christopher
Ricouer, Paul, 12
Robinson, Douglas, 234
Rogers, Richard A., essay on rhythm, 58
Romer, Marcus, 81
Rorty, Richard, 12
Royal Court Theatre, 197, 203
  adaptations of Chekhov, 174
  international work, 197
  Young Writers' Programme, 197
Royal Scottish Academy of Music and Drama, Glasgow, 141
Royal Shakespeare Company, 203
  and foreign work, 193
Rubin, David, 131, 132
Russo, Mariachiara, 74, 77

Sallenave, Danièle, 244
Sams, Jeremy, *Threepenny Opera*, 5, 126, 134
Scambiare arts festival, 141–2, 148
Scofield, Paul, 174
Scott, Clive, 66
Shakespeare, William, *Richard III*, 78
Sharrock, Thea, 207
Shaw, G.B., 188
Sheets-Johnstone, Maxine, 112
Shen Xiaomei, Madame Shen, 112–13
Shukman, Harold, 173, 174
Slater, Maya, translation of Maeterlinck's *Aveugles*, 44
Słobodzianek, Tadeusz, *Our Class*, 252
Snell-Hornby, Mary, 14
Sohoye, Ashmeed, 196
Solitude (Koltès), performance based translation
  analysis of the text, 50–9:
    alexandrine structures, 51–2, 54, 56, 58, 63; the Client, 53, 54, 56–7, 68–9; correlation, action and language, 56; the Dealer, 51, 52, 54, 55–6, 68–9; gaps and silences, 58–9; jazz comparison, 54, 59, 63; performativity, 53, 56, 57; rhythm and the use of sound, 54–7, 59; syntax as allegory of story, 57; syntax stretched to the limit, 51–3, 54, 56, 58; use of stichomythia, 57; verbal jousting, 51
  process and performer, 63–9:
    adapting Koltès use of metaphor, 67; Dalmasso performing French text, 64, 65, 66; rehearsal and evolution, 63–6; rhythmic structures of Hip Hop and slam/jazz, 63, 63–9; sound patterns, creative work on, 65, 67–8; speed of delivery, 64, 65; 'Stickman's performance, vocal and physical, 64, 65, 66, 69–70; strategy of sound over sense, 66–7
  the translation, 59–63: avoiding addition, 60–1, 66; choice of sounds, 60–1; complex French syntax, 60, 61–3, 66; retaining tempo, 60–1; text as musical score, 59; use of sound and rhythm, 6, 61, 62–3, 68–9
Solzhenitsyn, Alexander, 184
Sontag, Susan, 38, 40
Spellmeyer, Kurt, 12
Stafford-Clark, Max, *Letters to George*, 31
Stanislavski, 177
Stein, Lou, 206
'Stickman', David 'Stickman' Higgins, 64–9
Stoppard, Tom, 176

theatre translation, 1–2, 231–2
  as a function of *mise en scène*, 32–3
  in Germany, 126, 127
  practice and collaboration, 2–3, 7, 139
  and writing skill, 208
  see also translation
theatre translation practice
  comedy genre, 22–3
  contexts, 16–18
  critical versus creative, 12

theatre translation practice – *continued*
  cultural analysis, 17–18
  dangers of relocation of context, 21–8, 24–6
  dangers of semantic overload, 28
  double consciousness, 12–13
  interactive practice, 3, 28
  kinetics and kinesics, 26–8
  performability, 13–15, 18
  performable rhythms, 23–8
  performance led practice, 15
  practice and metaphor, 3–4, 18–21
  preservation of original cellular structure, 24
  reflective practice, 4, 11–13
  source and target texts, relationship, 20
  space and place as active participants, 17
  successful transformation, 28
The Threepenny Opera (Brecht), 5, 126–37
  critical response to, 128–9
  genesis of, 127–8: reinvigoration of lyrics, 134, 135; *The Moritat of Mack the Knife*, 128, 135
  Manchester Contact Theatre production (1994), 129
  National Theatre production (1986), 130
  National Theatre production (2002), 126, 130–7: adaptation of musical score, 130–1, 134; character of Macheath, 135; intimate staging, 131; 'Pirate Jenny's Song', 135–6, **136**; the Prologue, 131; translating 'social Gestus', 134–5; translating and staging, 133–7; up-dating with contemporary references, 132, 133–4
  original setting of play, 132–3
  Royal Court production (1956), 129
Thomas, Kristin Scott, 179
Tian, Min, 110
translatability, 21
translating Brecht, 126–7
translating Zapolska, 7, 254–65
  literary and personal reputation, 259
  modernising original speech register, 261–2
  politics of partitioned Poland, 256–8
  research methodologies, 254–6: first draft translation, 255; historiographical approach, 254; past performance and reception, 254–5, 263; post-performance re-fashioning, 255–6; research through practice, 255
  research performance of *The Morality of Mrs. Dulska*, 258–63: canonical status of work, 259; character of Mrs. Dulska, 258–60; Galicia contested setting, 262–5; use of Moniuszko's *The Haunted Manor*, 264–5
  Zapolska's corpus and its cultural transposition, 256–8: the bourgeoisie of partitioned Poland, 257–8; national and cultural identity, 256–7; the personal and the political, 258
  see also Poland; Polish theatre
translation
  Christopher Hampton, 183–4
  cultural and linguistic specificity, 243–4
  French language, 235
  literal translation, 6, 174, 177, 202, 206, 208–10, 222–3
  as a political act, 243–4
  resolution of critical versus creative dichotomy, 12–13
  theory, 12–13, 230–1, 237–8
translator
  as conduit, original text to ultimate spectator, 19
  as cultural promoter and mediator, 139
  responsibility of, 126
  role of, 136–7, 140
  social responsibility, 244
  status of, 184
translator-director *see* directing and translating
Traverse Theatre, Edinburgh, 149, 202–5, 208, 210

new and international work, 202, 203
  see also collaborative theatre translation projects
Trivedi, Harish, 232, 234
Trudeau, Pierre Elliott, 164–5
Tynan, Kenneth, 191

Ubersfeld, Ann, model of relationship between written and performance texts, 36, 40–3
University of Hawai'i Asian Theatre Program, 112–17, 120, 122
Upton, Carole-Anne, and Hale, Terry, 244

Vardi, Adva, 218
Veltruský, Jiří, 37
Venuti, Lawrence, 234, 236, 243, 244
Verga, *La Lupa*, 185
Villon, François, 128
Virdi, Harvey, 132
Vitez, Antoine, 32–3, 101, 242, 244
von Horváth, Ödön *see* Horváth, Ödön von

Wainwright, J., 63
Walser, Martin, 185
Walshe, David, 146, 148, 149
Walton, J. Michael, 127
Wannamaker, Sam, and Brecht, 129
Warchus, Matthew, 181
Weill, Kurt, 128, 130
  Weill estate, 134
Werber, Cassie, 231, 243, 245
  Chopped Logic theatre company, 231
Western Canada, bilingual drama, 6, 155–60
  audience understanding, 156–7
  code-switching vernacular, 6, 155–6
  francophone minorities, 155–8
  French language: in private life, 155, 160: theatre, 155–6
  L'UniThéâtre, 158
  public life, use of English, 155, 160
  use of 'Franglish', 158–9
  see also Cow-boy poétré
White, Douglas, 126
Whybrow, Graham, 195
Whyman, Erica, 207
Wichmann-Walczak, Elizabeth, 109
  translation of Chinese *jingju* plays, 113–16, **115**, 122
Wilder, Tim, 116
Williams, Roy, 196
  Sing Yer Heart Out for the Lads, 80–3
Williams, Tennessee, 192
Wollaston, Sam, 151
Worth, Irene, 175
Writernet, 6, 200

Yaari, Nurith, 217–18
Yan, Haiping, 110–11, 112, 121

Zapolska, Gabriela, 249–50
  Galicia, 257, 262–3
  and her contexts, 252–4
  Malka Szwarcenkopf (1897), 256
  Mrs Dulska in Court (1907), 262
  The Death of Felicjan Dulski (1907), 262
  The Morality of Mrs. Dulska (1906), 250; research performance, 258–63
  The Other (1898), 256
  as translator, 255
  Zapolska's Women: Three Plays, 250
Zatlin, Phyllis, 13–14
Znyk, Daniel, 238
Zuber, Ortrun, 237